CIVIL AVIATION AUTHORITY
LIBRARY
CLASS No. 69.003.2
COPY No. 1 (LTL)
AQUA GROUP

PRE-CONTRACT PRACTICE
FOR ARCHITECTS AND QUANTITY SURVEYORS

"For which of you, intending to build a tower, sitteth not down first, and counteth the cost, whether he have sufficient to finish it."
ST. LUKE XIV, 28

PRE-CONTRACT PRACTICE

FOR ARCHITECTS AND QUANTITY SURVEYORS

SEVENTH EDITION

THE AQUA GROUP

With sketches by
Brian Bagnall

BSP PROFESSIONAL BOOKS
OXFORD LONDON EDINBURGH
BOSTON MELBOURNE

Copyright © The Aqua Group 1960, 1964, 1967, 1971, 1974, 1980, 1986

All rights reserved. No part of this publication may be reproduced, stored in a retrieval system, or transmitted, in any form or by any means, electronic, mechanical, photocopying, recording or otherwise without the prior permission of the copyright owner.

First published in Great Britain by
  Crosby, Lockwood & Son Ltd 1960
Second edition 1964
Third edition 1967
Fourth edition 1971
Fifth edition published by
  Crosby Lockwood Staples 1974
Sixth edition published by
  Granada Publishing Ltd 1980
Seventh edition published by
  Collins Professional and Technical Books 1986
Reprinted by BSP Professional Books 1990

British Library
Cataloguing in Publication Data
  Pre-contract Practice: for Architects and
    Quantity Surveyors. — 7th ed.
  1. Building — Contracts and
    specifications
  I. Aqua Group
  692   TH425

ISBN 0–632–02986–2

BSP Professional Books
A division of Blackwell Scientific
  Publications Ltd
Editorial Offices:
Osney Mead, Oxford OX2 0EL
  (Orders: Tel. 0865 240201)
25 John Street, London WC1N 2BL
23 Ainslie Place, Edinburgh EH3 6AJ
3 Cambridge Center, Suite 208,
  Cambridge MA 02142, USA
54 University Street, Carlton,
  Victoria 3053, Australia

Set by Columns of Reading
Printed and bound in Great Britain by
Mackays of Chatham PLC, Chatham, Kent

CONTENTS

	Authors' note to the first edition	vi
	Authors' note to the seventh edition	vii
	Introduction	1
1	The First Stage . . .	3
2	Approximate Estimates, Cost Planning and Control	13
3	Drawings	31
4	Schedules	42
5	Specifications	50
6	Bills of Quantities	67
7	Sub-Contractors and Suppliers	78
8	Obtaining Tenders	89
	Index	99

AUTHORS' NOTE TO THE FIRST EDITION

At a meeting arranged by the Junior Organisation Quantity Surveyors' Committee of The Royal Institution of Chartered Surveyors, held at the Talbot Restaurant, London Wall, a number of architects were present, and following a talk by a well known builder on 'The Preparation of a Builder's Estimate', there was a lively discussion from which it appeared that many of the troubles which arose in building were due to inadequate pre-contract preparation.

It was suggested that if some of the younger members of our two professions felt that was the case, they should get together and do something about it.

In answer to that challenge a group of us dined together and talked about our work. From that beginning, *Pre-Contract Practice for Architects and Quantity Surveyors* has emerged.

For convenience, we adopted the name 'Aqua'. There is no significance in the choice of the word and, indeed, the reasons for choosing it are now lost in obscurity. For us, however, it will always be a reminder of many hours of discussion and hard work, a full measure of good humour, and much learned by members of two professions about each other's problems.

The sketches were drawn by Brian Bagnall.

>PETER JOHNSON, FRICS, FIArb. (Chairman)
>H. E. D. ADAMSON, FRIBA
>HARRY L. AGER, ARICS
>BRIAN BAGNALL, BArch. (L'pool)
>A. T. BRETT-JONES, ARICS
>F. S. JOHNSTONE. ARICS
>JOHN KEMP, ARIBA
>A. G. NISBET, BA (Arch.), FRIBA
>C. A. ROGER NORTON, AA Dipl. (HONS), FRIBA
>A. J. OAKES, FRICS, FIArb.

August 1959

AUTHORS' NOTE TO THE SEVENTH EDITION

It is now over a quarter of a century since the publication of the first of the Aqua Books – *Pre-Contract Practice*. Five of the original authors are still participating in producing this new edition – the seventh.

We have taken the opportunity to revise many of the examples and bring the text up to date where necessary.

Since the sixth edition was published, the JCT Intermediate Form of Contract (IFC 84) has become available and new systems of procurement have emerged. Efficient communications between the various parties is the key to good practice and in 1986 the Coordinating Committee for Project Information (CCPI) will publish its code for the Common Arrangement which will provide a new discipline for the industry in the flow of information.

We believe the principles set out in this book, whilst related to traditional competitive tendering using Bills of Quantities and the JCT Standard Form of Building Contract (JCT 80), continue to provide guidance to good practice. We believe this is so whatever form of contract, method of procurement, or information system is used.

This edition is being published concurrently with the new edition of *Contract Administration* – the sixth. These two books together with our book *Tenders and Contracts for Building* will, we hope, give comprehensive cover for those involved in the management of building projects. In addition, our book *Fire and Building*, published in 1984, not only gives cover for those involved with a fire but deals with the various measures of prevention through design and management.

We are indebted to Brian Bagnall for the sketches.

JAMES WILLIAMS, DA(Edin.), FRIBA (Chairman)
BRIAN BAGNALL, BArch. (L'pool)
TONY BRETT-JONES, CBE, FRICS, FCIArb.
PETER JOHNSON, FRICS, FCIArb.
FRANK JOHNSTONE, FRICS
ALFRED LESTER, Dip. Arch., RIBA
JOHN OAKES, FRICS, FCIArb.
QUENTIN PICKARD, BA, RIBA
GEOFFREY POOLE, FRIBA, ACIArb.
COLIN RICE, FRICS
JOHN TOWNSEND, FRICS, ACIArb.

"... to design ... down to the last ... doorknob ..."

INTRODUCTION

We have never defined the objects of the Aqua Group but our purpose has evolved over the years and we have come to see it as setting down in clear, concise and practical terms the principles of good practice in the work of the architect and the quantity surveyor.

Just as a building needs a sound foundation, so does professional practice associated with building and the latter must be based on the proper and conscientious application of established and proven principles.

We must not however lose sight of the fact that building procedures, including pre-contract work, are undergoing change and to a considerable extent it is now recognised that to design and specify a building down to the last cupboard door knob before tenders are invited may be neither necessary nor even desirable. The client's best interest may well be served by arranging for an early start with a view to achieving early completion with its accompanying quicker return on the capital invested. There is little objection to this provided certain fundamental rules are followed.

In the first place it is essential that the policy to be followed is decided at the inception of the project, so that the pre-contract programme and procedure can be drawn up accordingly.

Secondly the method of appointing the contractor, be it by single or two stage competitive tender, or by nomination, must be decided according to the requirements of the particular case.

Thirdly the form of contract to be used must be suitable for the procedure adopted.

Finally the design and specification of the work must always be sufficiently far ahead of construction as will enable the contractor properly to plan the work, obtain materials and build in an efficient and economic manner. It is essential that this is adhered to, not only in the pre-contract period, but also throughout the actual construction.

Another matter which has necessitated a major change in the pre-contract procedure is the trend towards bringing the contractor into the design process at an early stage. There is much to be said for this in cases where the contractor may make a positive contribution to good economic design, in cases, for instance, where the way in which plant can be utilised may have an influence on the design and cost, or where specialised or proprietary methods of construction are to be used.

It should always be borne in mind, however, that early introduction

PRE-CONTRACT PRACTICE

of the contractor should be prompted only by his being able to make a definite contribution to the project during the design stage. Its purpose is not to shift responsibility for design or preparation of working details from the architect to the contractor's staff. A similar situation can also exist in relation to certain types of sub-contractor's work, when the same rules must apply.

With sub-contractors, however, there may also be an essential element of design and in these cases the responsibility for design must be properly identified and the client must be properly safeguarded. We discuss this aspect in Chapter 7.

The use of these procedures has led to various new methods of selecting contractors and negotiating and arranging contracts. We refer to methods of obtaining tenders in Chapter 8 and the subject is dealt with more fully in our book. *Tenders and Contracts for Building.*

The current edition of the *Code of Procedure for Single Stage Selective Tendering*, published by the National Joint Consultative Committee for Building, was issued in December 1977. The principles which it sets out have not changed and the provisions for altering tenders after they have been opened, in order to correct errors, remain. We were inclined to criticise these provisions following the publication of the 1969 edition of the code as we felt that alternative 2, which allowed for tenders to be altered, was an undesirable expedient rather than good practice. We still consider that alternative 1, which does not permit the alteration of tenders, is to be preferred, but we acknowledge that tenderers are frequently permitted to correct genuine errors and it is appropriate, therefore, that the code should make provision for this to be done.

The 1980 edition of the *JCT Standard Form of Building Contract*, with its related documentation for domestic and nominated sub-contractors is now well established and the procedures imposed upon architects and quantity surveyors are dealt with in some detail in this book and in our book *Contract Administration.*

Our aim has always been to set out what we consider to be good practice. In doing so in this seventh edition we have considered it wise to continue to base our recommendations on the premise that we are dealing with a typical and reasonably normal building project, for which recognised competitive tendering procedure is adopted, and for which the Standard Form of Building Contract with Quantities is used. Where appropriate, however, we have also commented briefly on good practice relating to other procedures.

If our recommendations are followed, a good standard of practice will be established. Modifications in procedure can be made to suit new circumstances and changing conditions without reducing the high standard of work which the architect and the quantity surveyor have a duty to provide.

Chapter 1

THE FIRST STAGE . . .

The Simon Report, published in 1944 – over 40 years ago, set out what still remains an excellent summary of good practice in pre-contract procedure. Ten stages were identified, as follows:

(1) The architect, with the building owner, or employer, as he is known in standard forms of contract today, prepared a statement of the latter's requirements (the design brief).
(2) He selected the consultants and the quantity surveyor, obtaining where necessary, the approval of the employer.
(3) He prepared sketch plans and, with the quantity surveyor, a preliminary approximate cost.
(4) He selected such sub-contractors as he considered it necessary to nominate in advance of the signing of the contract.
(5) With the help of the consultants and nominated sub-contractors, he prepared a full set of drawings and specifications.
(6) The quantity surveyor prepared detailed bills of quantities, describing in words every service to be performed.
(7) Tenders were obtained on the basis of the bills of quantities from a selected list of firms.
(8) The architect, with the employer, selected the general contractor.
(9) He, with the consultants and quantity surveyor, prepared the contract documents.
(10) The general contractor signed the main contract and contracts with the sub-contractors – some nominated by the architect, some selected by himself.

Reading this through modern eyes, one can immediately detect a different emphasis from that which prevails today, but it would be difficult to improve on it as a summary of basic good practice, except perhaps to incorporate into it the processes of cost planning which have now become an essential feature of pre-contract procedure. It is also immediately obvious that no mention is made of planning matters and the other numerous statutory requirements which affect the design of the building.

The various stages of this pre-contract period are set out in a different form in the Outline Plan of Work in diagram 1 of the *Handbook of*

Outline plan of work
Plan of work diagram 1

Stage	Purpose of work and decisions to be reached	Tasks to be done	People directly involved	Usual terminology
A. Inception	To prepare general outline of requirements and plan future action.	Set up client organisation for briefing. Consider requirements, appoint architect.	All client interests, architect.	Briefing
B. Feasibility	To provide the client with an appraisal and recommendation in order that he may determine the form in which the project is to proceed, ensuring that it is feasible, functionally and financially.	Carry out studies of user requirements, site conditions, planning, design, and costs, etc., as necessary to reach decisions.	Clients' representatives, architects, engineers, and QS according to nature of project.	
C. Outline Proposals	To determine general approach to layout, design and construction in order to obtain authoritative approval of the client on the outline proposals and accompanying report.	Develop the brief further. Carry out studies on user requirements, technical problems, planning, design and costs, as necessary to reach decisions.	All client interests, architects, engineers, QS and specialists as required.	Sketch Plans
D. Scheme Design	To complete the brief and decide on particular proposals, including planning arrangement appearance, constructional method, outline specification, and costs, and to obtain all approvals.	Final development of the brief, full design of the project by architect, preliminary design by engineers, preparation of cost plan and full explanatory report. Submission of proposals for all approvals.	All client interests, architects, engineers. QS and specialists and all statutory and other approving authorities.	

Brief should not be modified after this point.

			Working Drawings
E. Detail Design	To obtain final decision on every matter related to design, specification, construction and cost.	Full design of every part and component of the building by collaboration of all concerned. Complete cost checking of designs.	Architects, QS, engineers and specialists, contractor (if appointed).

Any further change in location, size, shape, or cost after this time will result in abortive work.

F. Production Information	To prepare production information and make final detailed decisions to carry out work.	Preparation of final production information i.e. drawings, schedules and specifications.	Architects, engineers and specialists, contractor (if appointed).
G. Bills of Quantities	To prepare and complete all information and arrangements for obtaining tender.	Preparation of Bills of Quantities and tender documents.	Architects, QS, contractor (if appointed).
H. Tender Action	Action as recommended in NJCC *Code of Procedure for Single Stage Selective Tendering* 1977.	Action as recommended in NJCC *Code of Procedure for Single Stage Selective Tendering* 1977.	Architects, QS, engineers, contractor, client

Reproduced by permission of RIBA Publications Ltd.
Note: On small jobs without quantities Stage G will be omitted and Stages E and F may be combined.

PRE-CONTRACT PRACTICE

Architectural Practice and the *Architect's Job Book*, published by the RIBA. The relevant part of this is reproduced on pages 4 and 5 and it will be seen that this also provides a useful checklist of good practice.

We do not propose to prepare yet another summary of good practice but we will elaborate upon some of the matters referred to and endeavour to set out in some detail a guide to good pre-contract procedure.

The procedures described in the plan of work hold good no matter how the project is developed. However, architects and quantity surveyors are playing an increasing role in the actual initiation of schemes; identifying sites, bringing together interested parties and, in the commercial field, involving themselves in the financial aspects of the development. In such circumstances the consultants making up the design team may have come together and be acting as a steering group before the primary client is found. Again, in the commercial field, there may well be not one, but a group of clients, sharing investment in the project, with secondary clients waiting in the background as ultimate purchasers or tenants. Most commonly though, the formal appointment of the architect by a single client to prepare preliminary designs triggers the commencement of the pre-contract programme. It is the normal evolution of the brief and appointment of professional advisers which we will now discuss.

The Design Brief

The initial design brief is likely to be no more than a statement of intent made at the time of appointment of the architect and it must be developed into a full schedule of accommodation covering both technical and financial matters, and user requirements. Its development is, certainly in major projects, a team affair requiring the preparation of feasibility studies and cost reports and consultations with planning and other relevant authorities.

Regular meetings will be required with the client and all consultants to exchange information, discuss findings and receive further instructions. All such meetings should be minuted and this applies equally to small and large jobs.

The client should be made aware that once Scheme Design (Stage D in the Outline Plan of Work) is complete, the brief should not be modified without the client being alerted to the possibility of delays in the overall project programme and consequential additional fees in respect of abortive work.

Appointment of the Design Team

The earliest possible appointment of all the professional advisers who

THE FIRST STAGE...

make up the design team will ensure the most efficient development of the brief. Often the architect, as the first appointed, will be called upon to advise the client on the appointment of the other consultants. Having given the advice, he should ensure that the actual appointments are made by the client direct and not by himself, otherwise he may be deemed to be liable for their professional performance.

All appointments should be in writing. The responsibilities of each party, including the fees to be paid and professional indemnity insurance cover required, should be clearly set out and understood by the others. Standard forms of appointment are published by most professional bodies and these may be used with advantage. These standard forms are often backed up by booklets giving a detailed description of the services provided and these might well be presented to a client who has not previously employed such consultants.

Any form of agreement with a public authority or large corporation should be signed by an authorised officer, or may be sealed.

Funding

Before the professional advisers become involved in detailed work, the means of payment for the project should be declared. This is usually called the 'funding arrangement'. The source of finance will vary according to the client or agency authorising the work and in many instances the methods of funding will have an effect on the overall tender and construction programme, perhaps even on the design of the building. Consequently this matter should be discussed at an early stage.

Sources of funding include:

- Private means of an individual, company, or building society
- Central government
- Local authority
- Pensions funds
- Institutions
- Banks
- Pre-let arrangements

Two common situations serve to illustrate how funding arrangements can have a direct effect on the building programme. Central and local government authorities normally allocate funds to be spent in specific financial years and the construction must be planned to ensure that the expenditure is incurred during the periods for which the necessary money has been budgeted. This may involve careful phasing of the work over two or more financial years. In a commercial situation there might be a pre-let arrangement in which an agreement to lease states specific times for occupation of various sections of a building complex,

PRE-CONTRACT PRACTICE

tied in with capital payments from the lessee to the developer. Here again the funding arrangements become essential to the planning, tender and construction programmes.

Pre-contract Programme

As soon as possible after his appointment, the architect should prepare and present to the client for his approval a programme of activities leading up to the commencement of the contract.

Consideration of the scheme by government departments and local authorities may take a long time and the architect should not underestimate the period needed for this. The employer may be a corporate body and managers or departmental heads will probably have to be consulted at various stages and will also need time to study drawings and reports. In short, as much care must be exercised in forecasting time when the drawings are out of the office for consultation or approval as when they are in course of actual preparation.

The programme should allow adequate time for the preparation of a full set of drawings and specification notes by the architect and the consultants, bearing in mind that none can complete his work without taking the proposals of the others into account. It should allow time for particulars of sub-contracts to be prepared, time for the preparation of bills of quantities and adequate time for the contractors to prepare their tenders. After receipt of tenders, time must be allowed for them to be examined, for the priced bills of quantities to be checked and for a report to be made to the employer. Finally, though this goes beyond the scope of our present deliberations, further time must be allowed between the acceptance of a tender and the actual start of work for completion of the contract documents and for the contractor to plan the work and organise his site and office arrangements.

If the programme has been properly thought out and adhered to by all concerned, it will pave the way to efficient, speedy and economic building. It is important that progress in relation to the overall programme is carefully monitored and that any slippage is reported to the employer.

The pre-contract programme may be set out in network form, analysing graphically the sequence and interdependence of the pre-contract operations. The more complex the building, the more helpful such a network will be.

Networks, however, especially the more complex ones, do not provide a quick visual programme suitable for general use in a busy office and tend to be ignored by the harassed draughtsman or taker-off. It is advisable, therefore, to prepare from the network a simplified programme in bar chart form as indicated in example A. This has the advantage of being speedily understood by everyone concerned with the

THE FIRST STAGE...

project. There are, of course, many options in layout and content and the example has been chosen as being simple, clear and adaptable.

In line with other industries, building has to keep pace with modern thought and techniques and computers are being used by construction companies as well as being applied to the design process, drawings and planning.

Such techniques should not be adopted just because they are fashionable, however; they should only be used when greater efficiency and productivity result.

Design Co-ordination

The communication of ideas and requirements among the several members of the design team becomes essential as the techniques of building, and legislative requirements, become more complicated. The more services there are in a building, the greater the need to consider their coordination. It is for the design team to look critically at communications between each member at the very onset of a scheme to ensure that each requirement is reflected in the design at the proper time. Lack of co-ordination can lead to abortive design work and may result in unsatisfactory design solutions.

"Building is having to keep pace with modern thought and techniques..."

EXAMPLE A   PRE-CONTRACT PROGRAMME

THE FIRST STAGE . . .

Tendering Procedure and Contractual Arrangements

Depending on the client's brief and the time available for the total design and building process, there may be a number of different ways of dealing with the design, tendering and contractual arrangements. Our book *Tenders and Contracts for Building* examines in detail the various options and the contracts appropriate thereto. It is important that a decision on the method to be used, according to the circumstances, be made at a very early stage.

We do not propose to examine the options available here, but it may be helpful to summarise the standard forms of contract prepared by the Joint Contracts Tribunal which are available to suit varying situations. These are:

- JCT 80 – Standard Form of Building Contract 1980, which is published in separate local authority and private editions, each with quantities, with approximate quantities and without quantities.
- IFC 84 – JCT Intermediate Form of Building Contract 1984.
- Agreement for Minor Building Works 1980
- CD/81 – JCT Standard Form of Building Contract with Contractor's Design 1981.

The JCT has also issued in Practice Note 20 guidance on deciding on the appropriate form of JCT main contract to be used in various circumstances.

Incomplete Information

If it is not possible to allow sufficient time to complete the detailed design before the building work begins, it may be necessary to use bills of approximate quantities for all or part of the scheme; to appoint specialist sub-contractors in advance of the main contractor; to negotiate the tender price; in fact, to start building work on a limited amount of information. This has been recognised by the JCT, who have introduced a version of the Standard Form of Building Contract where bills of approximate quantities are used. Where this occurs the architect, other consultants and the quantity surveyor must have in mind that outstanding information must be provided without delaying progress on site, that any approximate elements must be remeasured and that greater cost control during building is essential.

This imposes on the design team the need for a realistic understanding of the building process. A schedule of outstanding information should be drawn up, indicating the date by which each item is required, and every effort should be made to adhere to it, so that delays are avoided.

PRE-CONTRACT PRACTICE

The Sketch Scheme

The first key event in the pre-contract programme is the approval by the building owner of the finalised sketch scheme which should consist of:
- Sketch drawings of a type readily understandable to the layman.
- An explanation of the drawings and their relationship to the design brief.
- A description of the materials and methods of construction.
- A description of the services and mechanical plant.
- Details of site works.
- A note of what statutory consents, even if only in principle, have been obtained.
- An approximate estimate of cost including professional fees and VAT and, if required, a cost plan.

Chapter 2

APPROXIMATE ESTIMATES, COST PLANNING AND CONTROL

Care should be taken in reporting of estimates to indicate the degree of approximation. The basis on which a particular estimate is made should be agreed with the employer as regards content: what it should include (or, equally important, exclude) and the date on which the estimate is to be based (i.e. whether it should relate to a 'base' date, the 'tender' date, or be the actual estimate of final cost). This is particularly important during periods of rapid inflation.

The terms 'cost planning' and 'cost control' are widely used in connection with building work but it must be appreciated that the word 'cost' applies to the cost of the work for the employer. 'Price' would perhaps be a more accurate term, but in this chapter we have adhered to the word 'cost'.

In building, cost is always important and the occasions on which the employer gives the architect a free hand to design a building regardless of cost are so few and far between that they can be ignored when considering the normal work of the architect and the quantity surveyor. Indeed, so important are the financial aspects of building that the reckoning and controlling of the expenditure on any project must be regarded as fundamental parts of the design process. Practice in these matters will vary considerably according to the type of job and the employer's requirements. The high costs of running and maintaining buildings may well require in-depth studies of 'cost in use' related to initial building cost.

There are three basic operations involved:

- Preparation of approximate estimates giving estimates of the cost of the whole or part of the project at various stages before tenders are invited.
- Preparation of a cost plan which apportions the estimated expenditure among the various elements of the building, enabling those concerned with the design to know how the expenditure is being incurred; where, if additional expenditure is considered desirable on one aspect of the building, economies can be effected elsewhere; and generally to enable the architect and the quantity surveyor working together to ensure that the employer is getting the best value for his money.

"... design a building regardless of cost ..."

- The exercise of cost control by the regular checking of the cost plan as the design is developed, thus ensuring that changes in the proposed expenditure are brought to light at an early stage, enabling modifications to be made elsewhere in the design if necessary.

The integration of cost into the design process is illustrated in the diagram opposite.

The procedures to be followed in approximate estimating and cost planning will be governed to a large extent by the employer's attitude to the financial aspects of the proposed development. In most cases the approach to the finances of a project comes under one of three headings. These are:

- Cost limit or fixed capital expenditure. In such a case the employer will state in his initial brief to the architect how much he is prepared to spend and will require the most suitable building he can obtain for that sum.
- Unit cost limits. These apply usually in the case of public bodies or other employers responsible for much repetitive building work, standard cost limits being applied on a unit basis, such as £x per person accommodated or £x per square metre of floor area. The best building must then be provided within these limits.

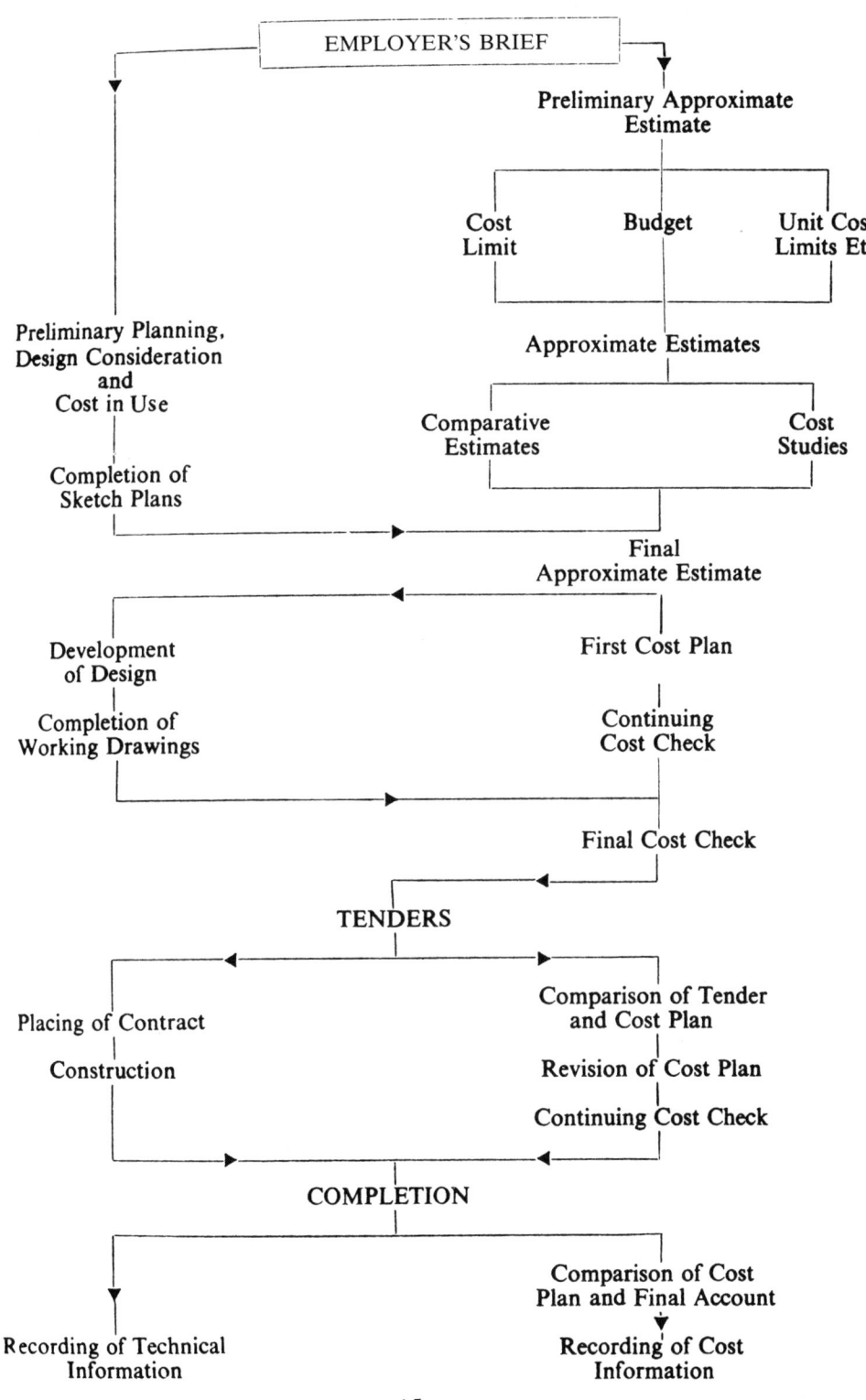

- Budget, based on estimated cost of proposed development. This is probably the most common approach in the case of the private developer. In the first instance he obtains guidance on costs from the architect and quantity surveyor from which he can make his first decisions regarding the brief to the architect and establish the cost limit for the job. This may involve him in considering such matters as the availability of the capital required, the likely return on the capital as an investment, or the economics of the proposed development in relation to his business.

From the foregoing it can be seen how important in any project is the accuracy of the approximate estimate given in the early stages and also the adequacy of cost planning and control during the pre-contract period.

Types of Approximate Estimate

The accuracy of approximate estimates must depend to some extent on the information on which they are based. It is therefore useful to classify the various types of approximate estimate since the information available to the quantity surveyor for each will vary. The classification set out below is not rigid and to a certain extent the four types of estimate listed overlap:

(1) *Preliminary approximate estimate*
This estimate may be derived from a rough floor area, cost per head or other similar unit method based on previous experience, taking into account site conditions and general cost trends in the industry. This will be the earliest estimate given and is only an indication to ensure that the project is of the same financial order as that contemplated by the employer. It should be remembered, however, that of all the cost information given to an employer during the planning of a building, the figures given in the first approximate estimate are the ones that remain most clearly in his mind. Furthermore it is frequently upon these figures that the whole economics of a development are based.

(2) *Approximate estimate based on floor area with appropriate parts taken from approximate quantities*
It will be more accurate than (1) and may be used as confirmation or otherwise of a previous estimate prepared when less information was available. If information is taken from a similar job it may be as accurate an estimate as is required before tenders. In this form of estimate only the building should be based on floor area. External works should

APPROXIMATE ESTIMATES, COST PLANNING AND CONTROL

be based on approximate quantities and any special or unusual items should be measured out. Specialist services should also be dealt with separately. This approximate estimate may be used for establishing a total cost which is not to be exceeded.

(3) *Approximate estimate based on approximate quantities*
With certain types of work, especially alteration work or unusual types of construction, or when relevant information from a similar job is not available, an approximate estimate must be prepared on this basis rather than (2) above. In any case it is necessary to prepare an estimate of this kind if effective cost control is to be exercised.

(4) *Approximate estimate from pricing an accurate bill of quantities*
This will give the most accurate approximate estimate possible but will normally only be available just before tenders. It is valuable as a check on the lowest tender and particularly useful where a small number of tenders is asked for.

Information required for approximate estimating

Set out below is a summary of the information which the quantity surveyor requires in order to prepare the various types of approximate estimate:

(1) *Preliminary approximate estimate*
In this case the estimate may have to be given without any drawings; nevertheless whether simple sketch plans are available or not the essential information required is as follows:
Type of building and its use.
The total floor area, overall height and number of floors.
The site and its nature.
An indication of the quality of work to be specified.
A brief outline of the engineering services.

(2) *Approximate estimate based on floor area*
Sketch plans:
1:200 or 1:100 Plan of each floor.
1:200 or 1:100 Elevations of most faces.
1:200 or 1:100 Sections.
1:500 or 1:200 Site plan showing extent of external works.
Specification notes:
Type of structure.
Materials for walls, floors and roof.
Floor loading.
Column spacing (if not shown on drawing).
General standard of finishings to walls, floors and roof.
Types of windows and doors.
Type of stairs.

PRE-CONTRACT PRACTICE

     Type of foundations.
     Extent of joinery fittings required.
     Type and scope of engineering and specialist services, including:
         Heating and hot water installation.
         Electrical installation.
         Mechanical ventilation.
         Sprinkler system.
         Lifts and escalators.
         Other services such as gas, vacuum, compressed air, etc.
     External services including positions of connections to all service mains and sewers.
     External works including types of road or paving, boundary walls, fences, gates etc.
     Any unusual conditions of work or contract.
(3) *Approximate estimate based on approximate quantities*
     Sketch plans or preliminary working drawings:
         1:100 Plans of each floor.
         1:100 Elevations of each face.
         1:100 Sections.
         1:500 or 1:100 Site plan showing extent of external works.
         1:20 Typical sections through building.
     Typical details of important features such as eaves.
     Specification notes: All the information listed in (2).
(4) *Approximate estimate from pricing accurate bill of quantities.*
     The bill of quantities and all information given to contractors tendering.

In all types of approximate estimate it must be borne in mind that, if works of alteration or extension are involved, the works will be subject to the addition of VAT.

After the preliminary approximate estimate various general options may be made the subject of cost studies and comparative approximate estimates may be prepared. These might deal with such matters as alternative shapes of plan, number of storeys, types of structural frame and other matters fundamental to the design.

When all concerned are satisfied with the general manner in which the employer's requirements can be met, sketch plans and approximate estimates can be finalised and, as a prelude to the preparation of working drawings, the cost plan can be prepared. The latter may well have been started in conjunction with the cost studies previously referred to.

APPROXIMATE ESTIMATES, COST PLANNING AND CONTROL

Cost Planning

In order to prepare a cost plan it is first necessary to sub-divide the building into a series of elements, each of which can be separately valued within the overall framework of the approximate estimate. Choice of elements will be based on the nature of the building and on any particular services or features it contains. It is essential, however, to use the same basic elements wherever possible in order to maintain continuity in compiling cost information.

During the detail design stage of a project any number of alternative designs and specifications may be considered for each element. These can be valued for the purpose of comparison and to enable the architect to see how various design decisions may affect the cost and how he can make the best use of the employer's money. The value of each element may be based on cost information from previous comparable jobs or on approximate quantities.

Set out below is the check list of elements used in the *Standard Form of Cost Analysis* published by the Building Cost Information Service of the RICS – from whom a booklet, setting out the full definitions, is available.

---

1. **SUBSTRUCTURE**
   All work below underside of screed or where no screed exists to underside of lowest floor finish including damp-proof membrane, together with relevant excavations and foundations.

2. **SUPERSTRUCTURE**
   2.A. **Frame**
   Loadbearing framework of concrete, steel or timber. Main floor and roof beams, ties and roof trusses of framed buildings. Casing to stanchions and beams for structural or protective purposes.
   2.B. **Upper Floors**
   Upper floors, continuous access floors, balconies and structural screeds (access and private balconies each stated separately), suspended floors over or in basements stated separately.
   2.C. **Roof**
       2.C.1. **Roof structure**
       Construction, including eaves and verges, plates and ceiling joists, gable ends, internal walls and chimneys above plate level, parapet walls and balustrades.
       2.C.2. **Roof coverings**
       Roof screeds and finishings. Battening, felt, slating, tiling and the like. Flashings and trims. Insulation. Eaves and verge treatment.
       2.C.3. **Roof drainage**
       Gutters where not integral with roof structure, rainwater heads and roof outlets. (Rainwater downpipes to be included in 'Internal drainage' (5.C.1.).)

PRE-CONTRACT PRACTICE

2.C.4. Roof lights
Roof lights, opening gear, frame, kerbs and glazing.
Pavement lights.

2.D. Stairs
2.D.1. Stair structure
Construction of ramps, stairs and landings other than at floor levels.
Ladders.
Escape staircases.
2.D.2. Stair finishes
Finishes to treads, risers, landings (other than at floor levels), ramp surfaces, strings and soffits.
2.D.3. Stair balustrades and handrails
Balustrades and handrails to stairs, landings and stairwells.

2.E. External Walls
External enclosing walls including that to basements but excluding items included with 'Roof structure' (2.C.1.).
Chimneys forming part of external walls up to plate level.
Curtain walling, sheeting rails and cladding.
Vertical tanking.
Insulation.
Applied external finishes.

2.F. Windows and External Doors
2.F.1. Windows
Sashes, frames, linings and trim.
Ironmongery and glazing.
Shop fronts.
Lintels, cills, cavity damp-proof courses and work to reveals of openings.
2.F.2. External doors
Doors, fanlights and sidelights.
Frames, linings and trims.
Ironmongery and glazing.
Lintels, thresholds, cavity damp proof courses and work to reveals of openings.

2.G. Internal Walls and Partitions
Internal walls, partitions and insulation.
Chimneys forming part of internal walls up to plate level.
Screens, borrowed lights and glazing.
Moveable space-dividing partitions.
Internal balustrades excluding items included with 'Stair balustrades and handrails'(2.D.3.).

2.H. Internal Doors
Doors, fanlights and sidelights.
Sliding and folding doors.
Hatches.
Frames, linings and trims.
Ironmongery and glazing.
Lintels, thresholds and work to reveals of openings.

3. INTERNAL FINISHES
3.A. Wall Finishes
Preparatory work and finishes to surfaces of walls internally.
Picture, dado and similar rails.

APPROXIMATE ESTIMATES, COST PLANNING AND CONTROL

3.B. Floor Finishes
Preparatory work, screed, skirtings and finishes to floor surfaces excluding items included with 'Stair finishes' (2.D.2.), and structural screeds included with 'Upper floors' (2.B.).

3.C. Ceiling Finishes

3.C.1. Finishes to ceilings
Preparatory work and finishes to surface of soffits, excluding items included with 'Stair finishes' (2.D.2.) but including sides and soffits of beams not forming part of a wall surface.
Cornices, coves.

3.C.2. Suspended ceilings
Construction and finishes of suspended ceilings.

4. FITTINGS AND FURNISHINGS

4.A. Fittings and Furnishings

4.A.1. Fittings, fixtures and furniture
Fixed and loose fittings and furniture including shelving, cupboards, wardrobes, benches, seating, counters and the like. Blinds, blind boxes, curtain tracks and pelmets. Blackboards, pin-up boards, notice boards, signs, lettering, mirrors and the like. Ironmongery.

4.A.2. Soft furnishings
Curtains, loose carpets or similar soft furnishing materials.

4.A.3. Works of art
Works of art if not included in a finishes element or elsewhere.

4.A.4. Equipment
Non-mechanical and non-electrical equipment related to the function or need of the building (e.g. gymnasia equipment).

5. SERVICES

5.A. Sanitary Appliances
Baths, basins, sinks, etc.
W.C.s, slop sinks, urinals and the like.
Toilet-roll holders, towel rails, etc.
Traps, waste fittings, overflows and taps as appropriate.

5.B. Services Equipment
Kitchen, laundry, hospital and dental equipment, and other specialist mechanical and electrical equipment related to the function of the building.

5.C. Disposal Installations

5.C.1 Internal drainage
Waste pipes to 'Sanitary appliances' (5.A.) and 'Services equipment' (5.B.).
Soil, anti-syphonage and ventilation pipes.
Rainwater downpipes.
Floor channels and gratings and drains in ground within buildings up to external face of external walls.

5.C.2. Refuse disposal
Refuse ducts, waste disposal (grinding) units, chutes and bins. Local incinerators and flues thereto.
Paper shredders and incinerators.

5.D. Water Installations

5.D.1. Mains supply
Incoming water main from external face of external wall at point of entry into building including valves, water meters, rising main to (but excluding) storage tanks and main taps.
Insulation.

PRE-CONTRACT PRACTICE

5.D.2. Cold water service
Storage tanks, pumps, pressure boosters, distribution pipework to sanitary appliances and to services equipment. Valves and tanks not included with 'Sanitary appliances' (5.A.) and/or 'Services equipment' (5.B.). Insulation.

5.D.3. Hot water service
Hot water and/or mixed water services.
Storage cylinders, pumps, calorifiers, instantaneous water heaters, distribution pipework to sanitary appliances and services equipment. Valves and taps not included with 'Sanitary appliances' (5.A.) and/or 'Services equipment' (5.B.). Insulation.

5.D.4. Steam and condensate
Steam distribution and condensate return pipework to and from services equipment within the building including all valves, fittings, etc. Insulation.

5.E. Heat Source
Boilers, mounting, firing equipment, pressurising equipment instrumentation and control, I.D. and F.D. fans, gantries, flues and chimneys, fuel conveyors and calorifiers. Cold and treated water supplies and tanks, fuel oil and/or gas supplies, storage tanks, etc., pipework (water or steam mains), pumps, valves and other equipment.
Insulation.

5.F. Space Heating and Air Treatment

5.F.1. Water and/or steam
Heat emission units (radiators, pipe coils, etc.), valves and fittings, instrumentation and control and distribution pipework from 'Heat source' (5.E.).

5.F.2. Ducted warm air
Ductwork, grilles, fans, filters, etc.
Instrumentation and control.

5.F.3. Electricity
Cable heating systems, off-peak heating systems, including storage radiators.

5.F.4. Local heating
Fireplaces (except flues), radiant heaters, small electrical or gas appliances, etc.

5.F.5. Other heating systems

5.F.6. Heating with ventilation (air treated locally)
Distribution pipework, ducting, grilles, heat emission units including heating calorifiers except those which are part of 'Heat source' (5.E.) instrumentation and control.

5.F.7. Heating with ventilation (air treated centrally)
All work as detailed under (5.F.6.) for system where air treated centrally.

5.F.8. Heating with cooling (air treated locally)
All work as detailed under (5.F.6.) including chilled water systems and/or cold or treated water feeds. The whole of the costs of the cooling plant and distribution pipework to local cooling units shall be shown separately.

5.F.9. Heating with cooling (air treated centrally)
All work detailed under (5.F.8) for system where air treated centrally.

5.G. Ventilating System
Mechanical ventilating system not incorporating heating or cooling installations including dust and fume extraction and fresh air injection, unit extract

APPROXIMATE ESTIMATES, COST PLANNING AND CONTROL

fans, rotating ventilators and instrumentation and controls.

5.H. Electrical Installations

5.H.1. Electric source and mains
All work from external face of building up to and including local distribution boards including main switchgear, main and sub-main cables, control gear, power factor correction equipment, stand-by equipment, earthing, etc.

5.H.2. Electric power supplies
All wiring, cables, conduits, switches, etc., from local distribution boards, etc., to and including outlet points for the following:
General purpose socket outlets.
Services equipment.
Disposal installations.
Water installations.
Heat source.
Space heating and air treatment.
Gas installation.
Lift and conveyor installations.
Protective installations.
Communication installations.
Special installations.

5.H.3. Electric lighting
All wiring, cables, conduits, switches, etc., from local distribution boards and fittings to and including outlet points.

5.H.4. Electric lighting fittings
Lighting fittings including fixing.
Where lighting fittings supplied direct by client, this should be stated.

5.I. Gas Installations
Town and natural gas services from meter or from point of entry where there is no individual meter; distribution pipework to appliances and equipment.

5.J. Lift and Conveyor Installations

5.J.1. Lifts and hoists
The complete installation including gantries, trolleys, blocks, hooks and ropes, downshop leads, pendant controls and electrical work from and including isolator.

5.J.2. Escalators
As detailed under 5.J.1.

5.J.3. Conveyors
As detailed under 5.J.1.

5.K. Protective Installations

5.K.1. Sprinkler installations
The complete sprinkler installation and CO_2 extinguishing system including tanks, control mechanism, etc.

5.K.2. Fire-fighting installations
Hosereels, hand extinguishers, fire blankets, water and sand buckets, foam inlets, dry risers (and wet risers where only serving fire-fighting equipment).

5.K.3 Lightning protection
The complete lightning protection installation from finials and conductor tapes, to and including earthing.

5.L. Communication Installations
The following installations shall be included:
Warning installations (fire and theft)
Burglar and security alarms.

PRE-CONTRACT PRACTICE

Fire alarms.
Visual and audio installations
Door signals.   Public address.
Timed signals.   Radio.
Call signals.   Television.
Clocks.   Pneumatic message system.
Telephones.

5.M. Special Installations
All other mechanical and/or electrical installations which have not been included elsewhere, e.g. Chemical gases; Medical gases; Vacuum cleaning; Window cleaning equipment and cradles; Compressed air; Treated water; Refrigerated stores.

5.N. Builder's Work in Connection with Services
Builder's work in connection with mechanical and electrical services.

5.O. Builder's Profit and Attendance on Services
Builder's profit and attendance in connection with mechanical and electrical services.

6. EXTERNAL WORKS
6.A. Site Works
6.A.1. Site preparation
Clearance and demolitions.
Preparatory earth works to form new contours.
6.A.2. Surface treatment
The cost of the following items shall be stated separately if possible:
Roads and associated footways.   Games courts.
Vehicle parks.   Retaining walls.
Paths and paved areas.   Land drainage.
Playing fields.   Landscape work.
Playgrounds.
6.A.3. Site enclosure and division
Gates and entrance.
Fencing, walling and hedges.
6.A.4. Fittings and furniture
Notice boards, flag poles, seats, signs.
6.B. Drainage
Surface water drainage.
Foul drainage.
Sewage treatment.
6.C. External Services
6.C.1. Water mains
Main from existing supply up to external face of building.
6.C.2. Fire mains
Main from existing supply up to external face of building; fire hydrants.
6.C.3. Heating mains
Main from existing supply or heat source up to external face of building.
6.C.4 Gas mains
Main from existing supply up to external face of building.
6.C.5 Electric mains
Main from existing supply up to external face of building.
6.C.6. Site lighting
Distribution, fittings and equipment.

APPROXIMATE ESTIMATES, COST PLANNING AND CONTROL

    6.C.7. Other mains and services
        Mains relating to other service installations.
    6.C.8. Builder's work in connection with external services
        Builder's work in connection with external mechanical and electrical services, e.g. pits, trenches, ducts, etc.
    6.C.9. Builder's profit and attendance on external mechanical and electrical services
6.D. Minor Building Work
    6.D.1. Ancillary buildings
        Separate minor buildings such as sub-stations, bicycle stores, horticultural buildings and the like, inclusive of local engineering services.
    6.D.2. Alterations to existing buildings
        Alterations and minor additions, shoring, repair and maintenance to existing buildings.

7. PRELIMINARIES

Cost Checks

As the design is developed and decisions are made the cost plan must be checked to ensure that such decisions will not adversely affect the intended expenditure. Where the value of any element in the cost plan is seriously altered by a decision taken at this stage it will be necessary to review the value of other elements in the cost plan.

Close liaison between architect, consulting engineers and quantity surveyor during this process will enable all concerned to be kept informed of any matters which might affect the cost of the building and to take such steps as may be necessary to ensure that the authorised expenditure is not exceeded.

In this book we are not dealing with events after the receipt of tenders, but it should be borne in mind that the process of cost checking should continue throughout the period of the contract when variations should be considered in the same way that design decisions are considered during the pre-contract period. To enable this to be done it is necessary to compare the tender with the cost plan and to revise the cost plan in the light of prices contained in the contract bills of quantities. On completion of the contract a similar comparison should be made of the cost plan and the final account so that all matters relating to the cost of the building can be properly recorded to provide detailed cost information for use on subsequent occasions.

Cost Planning Documentation

It will be appreciated that in the cost planning process and in exercising control in the pre-contract stage the closest liaison must be effected between designer and quantity surveyor. The processes of cost planning and cost checking are continuous and therefore a form of documentation which provides for a continual up-dating of the relevant cost information is important. We set out the following examples of typical documents:

Example B. An elemental cost summary sheet.
Example C. An amplified cost plan/check sheet.
Example D. An approximate estimate note sheet.

Quantity surveyors may find it helpful to have these produced in different colours.

It will be noted that in the elemental cost summary certain of the less frequently used specialist elements have been grouped together.

The sheets can, of course, be overprinted with the firm's name and address and such other references as may be required.

Such documentation will prove helpful as a reference for the quantity surveyor when producing tendering documents. An elemental breakdown of the specification is an invaluable help to the taker-off.

The quantity surveyor initiates the amplified cost plan but it will be noticed that there is a column for architect's revised requirements. These sheets can act as a valuable tool of communication between architect and quantity surveyor. Any revisions made on the amplified cost plan sheet are picked up on an estimate sheet to bring the particular element up to date.

Copies of the amplified cost plan sheets are given to the architect so that he continually has an up-to-date version of what is in the cost plan. These sheets will be a help if they are beside him in the later stages of design and during the preparation of working drawings.

It is important that the quantity surveyor's master copy shall be carefully filed with a loose-leaf arrangement of insertions at any place, so that all work relating to a particular element can be filed under that element. This includes rough notes and notes of telephone calls which have a cost implication.

Cost Information

The value and success of approximate estimates and cost plans are dependent upon the accuracy and interpretation of the cost information used in their preparation. The most effective cost information is that derived from the analysis of the cost of work dealt with in one's own office, because only in these cases has one intimate knowledge of the job. Other sources of information useful for checking purposes are:

- Cost analyses published in the technical press.
- Cost analyses and cost studies published by the Building Cost Information Service of the RICS.

APPROXIMATE ESTIMATES, COST PLANNING AND CONTROL

- Cost limits and other cost information published by government departments for use in certain classes of public work.

However well prepared cost analyses may be, the manner in which they are used and interpreted is of the utmost importance. Comparison of cost analyses of different buildings of a similar type will show wide variations in both the overall cost and the cost of individual elements. Analyses therefore require the most careful examination and consideration in relation to the project in hand.

In addition other factors which will influence the cost of the project must be borne in mind. These include:

- Cost trends in the industry, both national and local.
- Changes in rates of wages, either agreed or under consideration.
- Changes in other elements of the cost of labour, e.g. holidays with pay, national insurance, tool money, travelling allowances and the like.
- Fluctuations in prices of materials.
- The influence of other work in the district on the availability of labour and materials.

These matters emphasise the need for care and discretion when making use of published cost information. Probably the most important factors in approximate estimating and cost planning are the experience and good sense of the quantity surveyor.

Computers

The extensive use of computers in quantity surveyors' offices opens the field for the development of computer-based costing systems. Computer hardware is being developed at an amazing pace and systems are now available, including appropriate software packages, at comparatively modest cost.

EXAMPLE B

Cost Plan/Check No. ....
Elemental Cost Summary

Job No. .... Job Title .... Date ....

ELEMENT	ESTIMATED COST £	ESTIMATED COST PER m²	UNIT COST	COMMENTS
1. Work below Lowest Floor Finish				
2. Frame				
3. Upper Floors				
4. Roof				
5. Rooflights				
6. Staircases				
7. External Walls				
8. Windows				
9. External Doors				
10. Internal Load-Bearing Walls				
11. Partitions				
12. Internal Doors				
13. Ironmongery				
14. Wall Finishes				
15. Floor Finishes				
16. Ceiling Finishes				
17. Decorations				
18. Fittings				
19. Sanitary Fittings				
20. Waste, Soil and Overflow Pipes				
21. Cold Water Services				
22. Hot Water Services				
23. Heating Services				
24. Ventilation & Air Conditioning				
25. Gas Services				
26. Electrical Services				
27. Special Services				
28. Drainage				
29. External Works				
30. Miscellaneous				
31. Preliminaries				
32. Contingencies				
CURRENT ESTIMATED COST				

Target Cost ....

Gross Floor Area ....

	Drawings Incorporated to Date				
Drwg. No.	Rev.	Date	Drwg. No.	Rev.	Date

EXAMPLE C

Amplified Cost Plan/Check No. ....

Job No. .... Job Title ....

Element No. .... Element Title ....

LOCATION	Specification of Work Included in this Element	Architect's Revised Requirements to be included in next Cost Check

Drawings Particularly Relevant to this Element

Drwg. No.	Rev.	Date	Drwg. No.	Rev.	Date

Cost Plan/Check No. ....

Estimated Cost of Element £ ....

Estimated Cost per m² of Element   ....

Date on which Cost Transferred to Elemental Cost Summary   _____

EXAMPLE D

Estimate Note No. _____

Job No. _____   Job Title _____

Element No. _____   Element Title _____

DATE	SUBJECT	COST EFFECT		*METHOD OF ASSESSMENT	SOURCE OF INFORMATION & FILING REF.
		Add	Omit		

			Incorporated in Cost Check No. _____
			Date _____
Nett Add/Omit	£		* e.g. 1. Quick Appraisal
Previous Element Cost B/F	£		2. Approx. Estimate
			3. Verbal Quote
			4. Written Quote
Total Element Cost Transferred	£		5. Accepted Quote
to Cost Check/Estimate Note No. _____			6. Final Account

Chapter 3

DRAWINGS

It goes without saying that drawings represent the most important means by which the designer conveys his intentions to the client, statutory authorities, quantity surveyor, contractor and sub-contractors, and their absolute clarity is essential. Clear, well-planned drawings not only make the information they contain easy to understand, they inspire confidence. On the other hand, poor drawings do little but reveal the designer's lack of knowledge and inability to conduct affairs in a well-ordered manner.

Recommendations on all matters relating to the preparation of drawings, especially working drawings, are contained in BS 1192: 1984. This should be studied carefully but the following notes may also be a useful guide to good practice.

Sizes and Layout of Drawing Sheets

It is now common practice for drawing offices to have pre-printed drawing sheets in a limited range of sizes. The obvious advantage is that time is saved in drawing out titles and routine information that must appear on all sheets. The source of the drawings can then be readily identified, kept in comprehensive sets and stored easily. An assortment of drawings of different shapes and sizes, with title panels and essential information in differing positions, is a cause of confusion.

In BS 3429, recommendations are made for drawing sheet sizes A0, A1, A2, A3 and A4. From the point of view of economy in printing standard sheets, this range might be limited to A1 with suitable title panels. These are certainly the most useful drawing sizes. A3 and A4 are suitable for schedules and occasional quick answer sketches.

As it is helpful to keep drawings and schedules in filing systems, it is obviously best to limit the number of sizes. In selecting sizes it is worth considering those which can be reproduced by available photocopying equipment, such as B4 or A3.

Every sheet should have a filing margin of 20 mm minimum width at the left hand edge to accommodate file punchings. It is also useful for pre-printed sheets to have a drawn margin round all sides. This provides horizontal and vertical datum lines for setting the sheet up on the board, helps contain the drawing area clear of future binding tape and acts as a tell-tale indicating if only part of a drawing has been issued to date.

Two types of title and information panel are recommended in BS 1192. The examples include space for CI/sfb references, but this will not normally be required unless a CI/sfb classified form of specification is being employed. On the other hand, as much space as possible should be provided for general notes and cross references which might be overlooked if put in the body of the drawing and for details of amendments.

The importance of adequate space for amendment notes cannot be over-emphasised. It is insufficient merely to add an amendment letter to the drawing number. Unless the amendments are adequately described and the date when they were made noted, it can be difficult for the quantity surveyor or contractor to discover changes made to the drawings.

Paper expands and contracts to some extent and it is therefore unwise to take dimensions off drawings by scaling. Even though the greatest care should be used to ensure that drawings are accurate the information panel should include a standard note that all dimensions should be read and not scaled. The space provided for indicating the scales should be of reasonable size as it is common for there to be drawings of more than one scale on a single sheet. It is of great value to have a drawn scale on every sheet, particularly since it has become common practice to store record drawings on microfilm. Misinterpretations of scale can have disastrous results when reference is made to stored information long after the building has been completed.

Although BS 1192 does not state any specific preference for pencil or ink line drawings, it does note that dimensions should be in ink. In addition it is to be recommended that notes and certainly titles and important references are also in ink. They are reproduced more clearly in microfilm. Prints exposed to prolonged site use are subject to considerable wear and fading and those reproduced from ink negatives tend to be clear for much longer.

The use of squared paper under the tracing sheet can be of great assistance in setting out drawings and achieving dimensional accuracy. It is also thought that such backing sheets enable even an indifferent draughtsman to produce quick free-hand drawings in ink. In fact this is most doubtful; most draughtsmen can rule lines far more quickly than they can draw them free-hand. There is also the risk of inaccuracy.

Scales

Difficulties arise if drawings are prepared to unusual scales. It is important to use scales which are in common use. The following are recommended:

DRAWINGS

- Location and block plans — 1:2500, 1:1250, 1:500
- Site plans — 1:500, 1:200
- General and location drawings — 1:200, 1:100, 1:50
- Component drawings — 1:100 (on schedules A3 and A4 size), 1:50, 1:20
- Details — 1:5, 1:1 (full size)
- Assembly drawings — 1:20, 1:10

Priority of Drawings

It must be remembered that a building consists of many elements: the structural frame, the walls, partitions and roof, the heating, lighting, ventilation, plumbing, and so on. All these elements form part of a whole. Without any one of them the building is incomplete and it therefore follows that the set of drawings which tell the man on the site how the building is to be constructed is not complete if any one of the many elements is not shown or catered for.

All these elements affect the architect's design, and are affected by it. Columns and beams have to be a certain size in certain places. Radiators and pipes have to be accommodated and require ducts and holes in floors and walls.

It must also be remembered that the only value of drawings to a scale of 1:50 or smaller is to show the principal outline and main dimensions of a building. Details of construction can seldom be designed or shown to a smaller scale than 1:20.

Quantity surveyors cannot properly prepare a bill of quantities from small scale drawings alone, neither can builders build properly from such drawings.

These few facts – known to all architects but forgotten by many – are of prime importance in determining the sequence in which drawings are prepared. To avoid confusion it is essential to plan the production of drawings. Reference to the RIBA Outline Plan of Work (see pages 4 and 5) will show how important this is, particularly through work stages E and F. It should also be borne in mind that in the sixth edition of the

PRE-CONTRACT PRACTICE

Standard Method of Measurement the General Rules set out details of the drawings required for the purpose of tendering (see Chapter 8).

The drawing sequence is therefore as follows:

(1) Outline plans and elevations (say 1:100) which, after receiving the client's approval, are sent to all consultants who will then prepare their draft schemes.
(2) Whilst (1) is in progress any services for which the architect may be responsible (e.g. plumbing, drainage, etc.) must be worked out in detail.
(3) Concurrently with (1) and (2) the design must be considered in detail and drafts of construction details prepared. It is important at this stage that finishes and materials should finally be decided as these can affect dimensions critical to the primary elements of the structure and consequently the work of other consultants.
(4) When the consultants' drawings are accepted, and not before, the assembly details (1:20 and larger if necessary), which have been drafted in outline, can be completed.
(5) Only now that all the detailed information has been assembled and co-ordinated can the final overall picture (originally the outline drawings) be completed with accuracy.
(6) Layout and site plans and sections are finally completed, incorporating information on all external services and the setting out of the buildings.

In the course of preparation of working drawings it is invaluable to prepare a schedule of all drawings and associated schedules. This is likely to be enlarged as the work proceeds but assists in the rational order of production and ensures a comprehensive system for numbering. It also advises the quantity surveyor and contractor of information yet to be prepared. A system of numbering should be devised that includes reference to the job number and a prefix indicating whether or not it is a working drawing (as opposed to a preliminary or design drawing).

Completed drawing schedules should be kept up to date by the insertion of the latest drawing revisions. These schedules can be issued to the quantity surveyor at the time of starting bills of quantities, and to all parties at the commencement of the contract and from time to time thereafter. There should then be no reason for any member of the building team to be working from superseded drawings.

Although it is difficult to achieve, the most important rule is to complete all drawings before the stage is reached when they are required by others. The more incomplete the drawings when bills of quantities are being prepared, the less accurate will be the bills and the more uncertain the employer's financial commitment. Subsequently, it is simply unrealistic to expect construction to proceed smoothly unless the contractor receives completed drawings well ahead of his programme requirements. To prepare drawings in a sequence other than that suggested above is to invite trouble in the form of a badly detailed building, constructional delays, a long list of claims for extras and finally a dissatisfied client.

Computer Aided Design

The introduction of CAD has many attractions to the architect and even apart from carrying out every conceivable calculation related to the building process, the range of drawing functions that can be performed is wide. The most simple will produce two-dimensional drawings containing no more information than is passed into the memory for each drawing. The most advanced will hold in memory a complete model of the building and on command reproduce drawings to any scale or projection of any component part, section or the whole of that building, amendments to the original model automatically appearing on all drawings as they are required and produced.

The use of such equipment does not change the ground rules for good practice, although the systemised approach that is necessary will improve the clarity of the information produced and make it more accessible. The standardisation of components, co-ordination of services and indeed all aspects of rationalisation, are greatly facilitated. The preparation of schedules, specifications and even bills of quantities can become merely an extension to the whole operation.

The cost of the necessary equipment, the programs, and of course the specialised operators, is very high. Investment can only be justified if the equipment is in constant use. Maintenance is a considerable problem, particularly as developments in the whole field are rapid and hardware quickly becomes obsolete.

CAD is a vast and specialist subject, and is mentioned only to suggest that the technique so far has not changed the essentials of good practice in communicating information.

Contents of Drawings

Survey Plan
(1) Existing site and surroundings.
(2) Positions of major natural features, trees, hedges, ponds, etc.
(3) Sufficient spot levels and contour lines, related to a specified datum, to enable a section to be drawn in any direction through that part of the site to be built over (including roads, drains, and

PRE-CONTRACT PRACTICE

   the like), accurate to within 100 mm at any given point.
   (4) Key plan.
   (5) Position, invert and cover or surface levels of existing drains and service mains.
   (6) Access to site for vehicles.
   Note: If the survey plan covers more than one standard sheet, a key showing relationships of the various sheets must be shown on each drawing.

Site Plan, Layout and Drainage All information from survey plan, plus:
   (1) Outline of buildings showing positions of entrance and all soil and surface water gullies and outgoing pipes.
   (2) Steps where they occur.
   (3) Floor levels, clearly indicated using same datum as for existing levels on survey plan.
   (4) New roads and paths with widths and levels marked.
   (5) Soil and surface water drains complete with pipe sizes, and connections to sewer. Clear distinction should be made between soil and surface water drains and manholes. (Manhole sizes, levels and invert levels should be shown on a separate schedule.)
   (6) Runs of gas, water and electric mains with levels and positions of:
       (a) connections to existing mains,
       (b) supply company's meters (external) and details of meter housings if required,
       (c) points of termination within buildings.
   (7) Indication of banking and cutting and areas of deposing and spreading surplus soil.
   (8) New levels of site in connection with the foregoing.
   (9) If site is undulating or steeply sloping, sections should be added to show principal areas of cutting and filling.
   (10) Details of fences.
   (11) Particulars of any new planting.
   (12) External lighting and road lighting.
   (13) Figured dimensions giving necessary details for setting out main lines of buildings on site.

DRAWINGS

Plans of All Floors for Services When outline plans to 1:100 scale are approved and settled, copy negatives should be taken of the drawings showing only door swings in addition to walls and partitions and without any dimensions or other notes. These can then be developed without confusion or loss of time to show:
(1) Electrical layout.
(2) Heating layout.
(3) Plumbing and internal drainage layouts.
(4) Gas layout.
(5) Sprinkler layout.
(6) Fire detection and protection equipment.
(7) Layout of any special services.

Foundation Plan
(1) Width and depth of all foundations to walls, piers, and stanchions, with levels to underside.
(2) Positions and levels of drains, gulleys and manholes close to foundations.
(3) Walls above dotted and with thickness figured.
(4) Figured dimensions to centre lines of all stanchions.
(5) Position of incoming service mains and service main ducts and trenches and their levels.

Plans of All Floors
(1) Complete plans through all openings at all floors and mezzanine floors.
(2) (a) Overall dimensions,
    (b) All external dimensions including all openings,
    (c) Internal dimensions so far as is necessary to establish positions of internal walls, partitions and fittings, and thickness of all partitions.
(3) Door swings.
(4) Fittings in outline only with reference to details.
(5) Numbers or letters of all doors, as BS 1192.
(6) Numbers or letters of all windows, as BS 1192.
(7) Names or numbers of all rooms and circulation spaces, see also BS 1192.
(8) Pipe ducts, vertical and horizontal, flues, mat sinkings and the like.
(9) Air bricks.
(10) Hatching to indicate materials of which walls and partitions are constructed.
(11) Numbers of stair treads.
(12) Floor finishes.

PRE-CONTRACT PRACTICE

Roof Plan
(1) Construction and slopes.
(2) Levels.
(3) Types of coverings.
(4) Falls.
(5) Rainwater outlets, gutters and pipes.
(6) Roof lights.
(7) Tankrooms, trap-doors, chimney stacks, vent pipes and other penetrations.
(8) Parapets, copings, balustrades.
(9) Duckboards, catwalks and escape stairs.
(10) Lightning conductors and flagpoles.

Elevations and Sections
(1) Elevations of all parts of the building.
(2) At least one longitudinal and one cross-section, with heights figured from datum. These should be carefully chosen to give the maximum useful information.
(3) New and old ground levels showing cut and fill.
(4) External materials.
(5) Windows with opening lights marked.
(6) Air bricks and vents.
(7) External plumbing.

Construction Details
(1) Sections through external walls, foundations and roofs.
(2) Plans, sections and elevations of all staircases.
(3) Lift wells.
(4) Any room or part of the building the setting out of which is difficult or which involves extensive fittings, fixtures, plumbing or special features, such as:
(a) kitchens,
(b) bathrooms,
(c) lavatories,
(d) special purpose rooms.
(5) Windows and doors.
(6) Part elevations of any part of the building containing special features, such as:
(a) entrances,
(b) special forms of construction,
(c) balconies,
(d) ornamental brickwork or stonework.
(7) Boiler rooms.
(8) Electric intake rooms.
(9) Vertical and horizontal pipe ducts.

DRAWINGS

(10) Radiator recesses.
(11) Fireplaces and flues.
Note: When preparing these drawings, it must be remembered that all forms of construction used in the building should be shown. Repetition should be avoided at all costs, for instance, if windows on plan and section are identical throughout, the plan and section should be shown once only. There is no virtue in repeating the detail and repetition gives rise to discrepancies.

Sections and plans should always be drawn through the most complicated parts of the building. As a general rule it is not necessary to draw sections through an entire building as this only results in repetition and a considerable waste of paper. Strip sections with adjacent elevations are preferable, and if possible all sections relating to one datum line, such as all those at ground floor level, should appear on one sheet.

Large Scale Details These comprise enlargements of component parts of assemblies which have been shown to 1:20 scale but which require a larger scale to show the full details.
(1) Cills, heads and jambs of windows and doors.
(2) String courses.
(3) Mullions and transoms.
(4) Timber sections such as handrails, window sections, joinery details.
(5) Jointing details of curtain walling and any specialised wall cladding.
(6) Kerbs.
(7) Staircases.
(8) Special fittings and fixtures.
(9) Any special feature which cannot be described or shown clearly to smaller scale.

Drawings for Records

Before starting work on site there are two stages at which it is important to keep records of issued drawings before they are subjected to any further amendment. These are:

- Drawings upon which the bills of quantities are based – a full set of drawings including schedules.

- Contract drawings – those to which the contract documents refer.

It is useful for these copies to be in negative form, so that prints can be obtained readily should any future disagreements arise. In due course these might be microfilmed for the records along with the 'as-built' drawings.

True and accurate 'as-built' drawings are of vital importance in the future maintenance of the building and copies should be issued to the employer upon completion of the contract. Their preparation depends on regular correction of the drawings as the work proceeds and as amendments are issued.

Co-ordinated Project Information

Having gone to some length to identify good practice in the preparation and presentation of drawings, and before we go on to consider schedules, specifications and bills of quantities, it is perhaps appropriate to mention the work that is being done to co-ordinate all forms of project information and documentation.

The RIBA, RICS, BEC and the ACE are the four sponsoring bodies for the Co-ordinating Committee for Project Information (CCPI) which is working for greater liaison among the various bodies and in the project information that their members produce. It is hoped that, in future, codes of procedure for drawings and specifications and the seventh edition of the Standard Method of Measurement (SMM 7) will be based on a common natural grouping of information. This should assist all members of the building team to communicate more effectively and enable the information produced during the whole process to be readily available to people when they need it – thus contributing to a more efficient and cost effective service to the employer.

DRAWINGS

"Quantity Surveyors cannot properly prepare a bill of quantities from small scale drawings alone"

Chapter 4
SCHEDULES

It is good practice to convey information for items, such as windows and ironmongery, by means of schedules. They provide a means of conveying the architect's wishes to other persons concerned – in particular, the quantity surveyor and, later, the builder – in a form which has the following advantages over what may be described as the literary or graphical methods:

- The checking of errors of omission or duplication is simplified.
- The counting of similar items for obtaining estimates or placing orders is simplified.
- The omission of information on an item is unlikely as the appropriate column in the schedule would then be blank.
- If the information is set down systematically prolonged searching through a specification is avoided.

The recording of information by the architect on a drawing or other document presupposes that someone will want to refer to it later; the simpler and more foolproof the task of reference can be made, the more rapidly and economically will the building grow.

At this point it may be well to define three terms used in this chapter in connection with schedules.

- ITEM is the thing described, it may be a door, a window, a manhole, or a complete room.
- SIZE includes all dimensions relevant to the particular schedule. Thus a door schedule will tabulate thickness, width and height of each door, but a finishes schedule will not record room dimensions.
- CHARACTERISTICS include all the information necessary to give a complete description of the item, such as quality of material, method of construction, and finish.

Schedules should be either the same size as the standard drawing sheet selected for the job, or, if smaller, bound together as a set in one of the accepted 'A' sizes.

Layout will vary according to the information to be conveyed. ITEMS may read downwards, as in a manhole schedule, with CHARACTERISTICS and SIZES across, or vice versa. The right-hand

SCHEDULES

"The recording of information . . . presupposes that someone will want to refer to it later . . ."

column should always be for NOTES as there is frequently some exception, reservation, or cross-reference to be recorded.

Once a full description or reference to a drawing has been made, it is pointless to repeat this for further similar items in the schedule. For example, having described or cross-referenced the construction of a particular type of door, it can be given a type number or letter and this only would be repeated subsequently.

CHARACTERISTICS, particularly on a finishes schedule, can often be defined by a code letter or number, together with a key, to eliminate constant repetition of descriptions. Any clear abbreviation, which will not be confused with others, can be used to save time, with, if necessary, an explanatory note in the last column at its first appearance. Thus on a finishes schedule 'KPS & 3' is readily recognised as 'knot, prime, stop and paint three coats of oil colour'.

Some information may more clearly be conveyed in a schedule by a dimensioned sketch.

The schedules are not intended to supplant the bill of quantities, which will contain the final and full description of the CHARACTERISTICS.

Examples of the following schedules are given at the end of this chapter:

(E) Window schedule, including ironmongery, glazing, frames and, in some cases, cills.
(F) Door schedule, including ironmongery, glazing, frames and thresholds.
(G) Finishes schedule, including small standardised fittings such as hat and coat racks.
(H) Decorations schedule.
(I) Manhole schedule.

This is not a comprehensive list; there are many other opportunities for using schedules.

The column headings under SIZES and CHARACTERISTICS selected for each schedule cannot be standardised as they must vary with different types of building, but the cardinal principle in selecting the column headings must be a constant regard for:

- Who is going to use the schedule?
- What information does he need?

If these rules are borne in mind, the documents will be comprehensive without being cluttered with gratuitous and irrelevant information.

It is our opinion that the way of presenting schedules shown in the examples which follow is more easily understood and less likely to lead to error in preparation or use than the presentation shown in BS 1192: 1984 (Construction Drawing Practice).

EXAMPLE E WINDOW SCHEDULE

WINDOW TYPE	A	B	C	D	E	ETC.
WINDOW REF.NO.	W1 01to10, 12to14, 22to24, 26to34 W2 01to10, 12to14, 22to24, 26to34	W1 15,21,35,36, 37,38 W2 15,21,35,36, 37,38	W1 11,25 W2 11,25	W1 16,20,36,37 W2 16,20,36,37	W1 17,18,19 W2 17,18,19	ETC.
NO.OFF	56	12	4	8		
OPENING SIZE	1575 x 900	1575 x 600	500 x 1575	2100 x 600		
WINDOW SIZE	1570 x 895	1570 x 595	495 x 1570	2095 x 595		
FIXINGS: HEAD	CONC.LINTOL WITH SINGLE BRICK CANT. CAST INTO FACE			IN-SITU CONC.		
JAMB	BRICKWORK					
CILL	SLATE WITH GALVANISED M.S. WEATHER BAR					
SUB-FRAMES	NIL	NIL	NIL	ANODISED ALUMINIUM		
MATERIALS	ANODISED ALUMINIUM			NIL		
BUTTS	NIL	NIL	SUPPLIED WITH FRAME	NIL		
PIVOTS	NIL	NIL	NIL	NIL		
FASTENERS	NIL	NIL	SUPPLIED WITH FRAME	SUPPLIED WITH FRAME		
STAYS	NIL	NIL	SUPPLIED WITH FRAME	SUPPLIED WITH FRAME		
OPERATING GEAR	NIL	NIL	'TELEFLEX' CAT.NO.X1577	NIL		
GLAZING	DOUBLE GLAZED UNIT	DOUBLE GLAZED UNIT	4mm CLEAR FLOAT	4mm CLEAR FLOAT 4mm LAMINATED		
FIXING	FACTORY GLAZED	FACTORY GLAZED	ANODISED ALUMINIUM BEADS	ANODISED ALUMINIUM BEADS		
NOTES	SASHES FITTED WITH MANUAL OVERIDE SAFETY STOP	SASHES FITTED WITH MANUAL OVERIDE SAFETY STOP		SEE DETAIL FOR JUNCTION WITH RETURN WINDOWS AT HIGH LEVEL: W1 16/17,20/19 W2 16/17,20/19		
DIAGRAMS	→ ←	→ ←				

general notes

revisions

project/client
SHOPS AND OFFICES
NEW BRIDGE STREET
BORCHESTER

drawing title
window schedule

number
456/9.2

scale   date 11.11.84   drawn GP   rev

Reed & Seymore
architects
12 The Broadway
Borchester BC4 2NW

EXAMPLE F   DOOR & IRONMONGERY SCHEDULE

	DA		DB	DC	DD	ETC.
DOOR TYPE NO.	DA		DB	DC	DD	
DOOR REF. NO.	1	2	1	1	1	
LOCATION	STAIRCASE LOBBY/OFFICES	STAIRCASE LOBBY/OFFICES	EAST STAIR	WEST STAIR	WASH ROOM	ETC.
TYPE DESCRIPTION	1 1/2 LEAF SINGLE SWING REBATED EDGE INTERNAL QUALITY PLY FACED SOLID CORE FLUSH PANEL – MAIN LEAF WITH GWPP GLAZED VIEWING PANEL.		INTERNAL QUALITY PLY FACED SOLID CORE FLUSH PANEL WITH GWPP GLAZED VIEWING PANEL.	INTERNAL QUALITY PLY FACED SOLID CORE FLUSH PANEL WITH GWPP GLAZED VIEWING PANEL	INTERNAL QUALITY PLY FACED SOLID CORE FLUSH PANEL	
FIRE RATING			30/30			
SIZE	(2040 x 826 x 12mm MEETING STILE (2040 x 526 x 46)		2040 x 907 x 46	2040 x 907 x 46	2040 x 826 x 46	
FINISH	POLYURETHANE SEAL	POLYURETHANE SEAL	POLYURETHANE SEAL	POLYURETHANE SEAL	PAINTED	
FRAME O/A SIZE & SECTION	2375 x 1770 FRAME 57 x 130 BOTTOM RAIL 45 x 100		2375 x 1770 REBATES 25 mm	2375 x 957 REBATES 25 mm	ETC.	
MATERIAL & FINISH	SOFT WOOD	--	PAINTED INTUMESCENT STRIPS TO DOOR REBATES			
SIDE LIGHT	1 NO. 6mm GWPP		1 NO. 6mm GWPP	1 NO. 6mm GWPP		
FAN LIGHT	1 NO. 6mm GWPP		1 NO. 6mm GWPP			
DIRECTION OF SWING	526 826	826 526	907 907	907		
BUTTS	1 1/2 PAIRS BUTTS – EACH LEAF					
LOCKS & LATCHES	::					
HANDLES	PULL HANDLE 826 LEAF	PULL HANDLE 826 LEAF	PULL HANDLE	PULL HANDLE		
KICK & PUSH PLATES	PUSH PLATES OPP. SIDE 826 LEAF KICKING PLATES BOTH SIDES BOTH LEAFS		PUSH PLATE OPPOSITE SIDE KICKING PLATES BOTH SIDES			
BOLTS	PR. FLUSH BOLTS TO 1/2 LEAF 526					
STOPS & STAYS	O/H DOOR CLOSER	O/H DOOR CLOSER	O/H DOOR CLOSER	O/H DOOR CLOSER		
MISCELLANEOUS	O/H DOOR CLOSERS TO LARGE LEAFS (526) ONLY					
DIAGRAMS.	DA		DB	DC	DD	

general notes

revisions

project/client: SHOPS AND OFFICES NEW BRIDGE STREET BORCHESTER

drawing title: door + ironmongery schedule

number: 456/13.2

scale | rev.
date 4.12.84 | drawn GP

Reed & Seymore architects
12 The Broadway
Borchester BC4 2NW

EXAMPLE G FINISHES SCHEDULE

general notes
1. FOR SUBSEQUENT DECORATIONS SEE DECORATIONS SCHEDULE

Floor	Room No	Room Name	Walls	Skirting	Floor Finish	Floor Screed	Ceiling	Window Cill	Notes
BASEMENT	B.1	LOBBY	F.F. BRICKWK.	100mm GRANO	20mm GRANO	30mm SCREED	PLASTER	NONE	
	B.2	SWITCH ROOM	PLASTER	"	"	"	"	"	
	B.3	BOILER ROOM	F.F. BRICKWK.	"	"	"	F.F. CONCRETE	"	SEE ENGINEER'S DRAWINGS FOR BASES ETC.
GRD. FLOOR	G.1	ENTRANCE/RECEPTION	PLASTER	19 x 100 SOFTWOOD	6mm CORK TILE	44mm SCREED	ACOUSTIC TILE	"	
	G.2	WASHROOM	"	75mm COVE TESS. TILE	12mm TESS. TILE	38mm SCREED	PLASTER	12mm BULL NOSE TILE	FOR FITTINGS SEE SANITARY FITTINGS
	G.3	STAIRS - WEST	"	19mm SFT. WD. TO DETAIL	CARPET	25mm SCREED	"	"	NON SLIP NOSING TO SPEC.
	G.4	STAIRS - EAST	"	GRANO	20mm GRANO	"	"	"	"
FIRST FLOOR	1.1	LOBBY	"	19/100 SOFTWOOD	CARPET	45mm SCREED	ACOUSTIC TILE	NONE	
	1.2	WASHROOM	"	75mm COVE TESS. TILE	12mm TESS. TILES	38mm SCREED	PLASTER	12mm BULL NOSE TILE	FOR FITTINGS SEE SANITARY FITTINGS
	1.3	OFFICE AREA	"	19 x 100 SOFTWOOD	CARPET	45mm SCREED	ACOUSTIG TILE	SLATE TO DETAIL	
	1.4	MANAGER	"	"	"	"	"	HARDWOOD	
	1.5	SECRETARY	"	"	"	"	"	"	
	1.6	INTERVIEW	"	"	"	"	"	"	
	1.7	STAIRS - WEST	"	19mm SFT. WD. TO DETAIL	"	25mm SCREED	PLASTER	12mm BULL NOSE TILE	
	1.8	STAIRS - EAST	"	GRANO	20mm GRANO	"	"	NONE	
SECOND FLOOR	2.1	LOBBY	"	19 x 100 SOFTWOOD	CARPET	45mm SCREED	ACOUSTIC TILE	"	
	2.2	WASHROOM	"	75mm COVE TESS. TILE	12mm TESS TILES	38mm SCREED	PLASTER	12mm BULL NOSE TILE	FOR FITTINGS SEE SANITARY FITTINGS
	2.3	OFFICE AREA	"	19 x 100 SOFTWOOD	CARPET	45mm SCREED	ACOUSTIC TILE	SLATE TO DETAIL	
	2.4	CONFERENCE	"	"	"	"	"	HARDWOOD	
		ETC.			ETC.			ETC.	

project/client: SHOPS AND OFFICES NEW BRIDGE STREET BORCHESTER

drawing title: finishes schedule

number: 456/17

date: 10.12.84   **drawn:** GP

Reed & Seymore architects
12 The Broadway
Borchester BC4 2NW

EXAMPLE H   DECORATIONS SCHEDULE

Reed & Seymore architects
12 The Broadway, Borchester BC4 2NW

project/client: SHOPS AND OFFICES, NEW BRIDGE STREET, BORCHESTER
drawing title: decorations schedule
number: 456/18
date: 12.12.84   drawn: GP

general notes
1. FOR COLOURS SEE DRAWING NO. 456/27

Floor	Room No	Room Name	Walls	Skirting	Floor	Doors	Door Frames/Linings	Windows	Ceiling	Notes
BASEMENT	B.1	LOBBY	NIL	NIL	NIL	3 COATS GLOSS FINISH	3 COATS GLOSS FINISH	NONE	2 COATS EMULSION	
BASEMENT	B.2	SWITCH ROOM	3 COATS GLOSS FINISH	"	"	"	"	NONE	"	
BASEMENT	B.3	BOILER ROOM	NIL	"	"	"	"	NONE	"	PIPEWORK DECS BY ENGINEERS
GRD. FLOOR	G.1	ENTRANCE/RECEPTION	VINYL FINISH PAPER	3 COATS GLOSS FINISH	POLYURETHANE SEAL	POLYURETHANE	"	SELF FINISH	SELF FINISH	
GRD. FLOOR	G.2	WASHROOM	3 COATS GLOSS FINISH	3 COATS GLOSS FINISH	NIL	"	"	3 COATS GLOSS FINISH	2 COATS EMULSION	
GRD. FLOOR	G.3	STAIRS	2 COATS EMULSION	3 COATS GLOSS FINISH	"	"	"	"	"	
GRD. FLOOR	G.4	STAIRS	"	NIL	"	"	"	"	"	
FIRST FLOOR	1.1	LOBBY	"	3 COATS GLOSS FINISH	"	"	"	NONE	SELF FINISH	NOTICE BOARD AS DR. FRAMES
FIRST FLOOR	1.2	WASHROOM	3 COATS GLOSS FINISH	NIL	"	"	"	3 COATS GLOSS FINISH	2 COATS EMULSION	
FIRST FLOOR	1.3	OFFICE AREA	2 COATS EMULSION	3 COATS GLOSS FINISH	"	"	"	SELF FINISH	SELF FINISH	
FIRST FLOOR	1.4	MANAGER	"	"	"	"	"	"	"	
FIRST FLOOR	1.5	SECRETARY	"	"	"	"	"	"	"	
FIRST FLOOR	1.6	INTERVIEW	HESSIAN	"	"	"	"	"	"	
FIRST FLOOR	1.7	STAIRS	2 COATS EMULSION	"	"	"	"	3 COATS GLOSS FINISH	2 COATS EMULSION	
FIRST FLOOR	1.8	STAIRS	"	NIL	"	"	"	SELF FINISH	SELF FINISH	
SECOND FLOOR	2.1	LOBBY	"	3 COATS GLOSS FINISH	"	"	"	NONE	SELF FINISH	NOTICE BOARD AS DR. FRAMES
SECOND FLOOR	2.2	WASHROOM	3 COATS GLOSS FINISH	NIL	"	"	"	3 COATS GLOSS FINISH	2 COATS EMULSION	
SECOND FLOOR	2.3	OFFICE AREA	2 COATS EMULSION	3 COATS GLOSS FINISH	"	"	"	SELF FINISH	SELF FINISH	
SECOND FLOOR	2.4	CONFERENCE	VINYL FINISH PAPER	"	"	"	"	"	"	
		ETC.			ETC.			ETC.		

EXAMPLE I   MANHOLE SCHEDULE

general notes

SEWER INVERT 5.791
CONCRETE BASE 150 THICK
BRICK SHAFT 600 x 450 WHERE DEPTH TO INVERT OVER 1500
STEP IRONS OVER 750 DEEP AT 450 CENTRES
ALL COVERS AS BS 497
S = STRAIGHT
C = CURVED
T = TAPER

revisions

project/client
SHOPS AND OFFICES
NEW BRIDGE STREET
BORCHESTER

drawing title

manhole schedule

number 456/20

scale — **date** 3.1.84 **drawn** GP **rev**

Reed & Seymore architects
12 The Broadway
Borchester BC4 2NW

No	FOR USE BY: Size Internally	Side	ARCHITECT Cover	Invert Level	Ground Level	Cover Level	QUANTITY SURVEYOR (filled in from drawings) Depth Excavation	Depth to Invert	Main Channel Size	Type	Branch Chnl One Side	Other Side	ARCHITECT Notes
1	1125 x 825	112.5	450 x 600 GRADE C FIG7	7315	8077	8460	0912	1145	100	C	4	2	
2	"	"	"	7214	8153	8382	1089	1168	100	S	1	4	
3	900 x 825	225	"	7020	8175	8175	1305	1155	100/150	C.T.	2	—	REVERSE INTERCEPTOR
4	"	"	450 x 600 GRADE C FIG8	6807	8080	8382	1423	1575	150	C	1	—	
5	900 x 675	"	"	6706	8080	8382	1524	1676	150	S	—	—	
6	"	"	"	6401	8062	8382	1811	1981	150	S	—	—	
7	"	"	450 x 600 GRADE C FIG5	6172	8001	8230	1979	2058	150/225	S.T	1x100 1x150	2	
8	"	"	"	6858	8077	8458	1369	1600	100	S	2	2	
9	1125 x 600	"	"	6629	7988	8458	1509	1829	100	S	4	1	
10	900 x 675	"	"	6477	7732	8382	1455	1905	100	C	—	1	
11	"	"	"	6330	7849	8306	1679	1976	100/150	S.T	2	—	
12	675 x 675	"	600 x 600 GRADE A FIG1	5867	7925	8230	2208	2363	225	S	—	1	FAI & INTERCEPTING TRAP
13	675 x 450	112.5	450 x 600 GRADE C FIG8	7398	8204	8384	0956	0994	100	C	—	1	
14	"	"	"	8255	8230	8534	0125	0279	100	C	—	—	

Chapter 5

SPECIFICATIONS

In the two previous chapters we have shown how, by preparing good drawings and schedules, the architect can convey most of the information about a proposed building first to the quantity surveyor and later to the builder. Nevertheless there is information which, by its nature, must be given by written description. For example, the quality of cement to be used or the method of laying granolithic paving cannot be shown graphically but must be described in writing.

The traditional document for conveying such information is the specification.

With the current emphasis on rehabilitation and refurbishment of existing buildings, an increasing number of smaller jobs are carried out using drawings and specification only, omitting the use of bills of quantities. In these circumstances the specification document will take on a wider role than merely defining the quality of the materials and workmanship and will be a contract document.

Whichever contract applies, with bills or without, the architect must be able to convey his intention, either to the quantity surveyor or to the builder, in a clear, logical and unambiguous manner.

With this in mind we have prepared a comprehensive set of specification headings, setting out, trade by trade, the information necessary either for the quantity surveyor to pass on to the builder in the bill of quantities or for the architect to pass to the builder in the specification.

The complete list of these headings is given below. The clause numbers in the Preliminaries section refer to the JCT 80 Standard Form of Building Contract.

The contractual significance of the specification varies according to the conditions of contract being used. Under JCT 80, with quantities, the specification as such is not a contract document and its essential requirements must be incorporated into the bills of quantities in the preliminaries, preambles and descriptions of the work generally. Alternatively it can be incorporated in the bills by direct reference to the document as a whole, but even if that is done it remains essential that the bill descriptions are still adequate to convey the designer's requirements without constant reference to the specification.

When the contract has been placed the architect must provide the

SPECIFICATIONS

contractor with two copies of the specification or, to use the wording of the contract, 'any description schedules or other like documents necessary for use in carrying out the works'.

"... the specification as such is not a contract document ..."

SPECIFICATION HEADINGS

A. Tendering Particulars
1. Name of Job.
2. Name, address, telephone number and reference of:
   (a) Employer.
   (b) Architect.
   (c) Quantity Surveyor.
   (d) Structural Engineer.
   (e) Service Engineer.
3. Tendering details:
   (a) Date to be sent out,
   (b) Date and time of submission,
   (c) Where tenders to be submitted.
4. List of contractors invited to tender.

B. Preliminaries
1. Address of site.
2. Place where all drawings may be inspected.
3. Access for contractors to inspect site.
4. Scope of contract.
5. Procedure; any special conditions or sequence of operations.
6. Particulars of access to site.
7. Form of contract to be used.
8. If standard form, any amendment in full.
9. List of drawings for preparation of bills of quantities.
10. Whether employer is or is not 'a contractor' under statutory tax deduction scheme (Recital 4 and Clause 31).
11. Scope of reference to arbitration (Article 5.1.6).
12. Date for completion (Clause 1.3).
13. Defect liability period (Clause 17.2).
14. Insurances:
    (a) Cover required under Clause 21.1.1.
    (b) Amount of indemnity (if any) required under Clause 21.2 and provisional sum in respect of same.
15. Liability for fire insurance (Clause 22).
16. Percentage to cover professional fees (Clause 22A).
17. Date for possession (Clause 23.1).
18. Liquidated and ascertained damages (Clause 24.2)
19. Period of delay (Clause 28.1.3):
    (a) By contingencies referred to in Clause 22.
    (b) For other reasons.
20. Period of interim certificates (Clause 30.1.3).
21. Retention percentage (Clause 30.4.1.1.).
22. Period of final measurement (Clause 30.6.1.2).
23. Period for issue of final certificate (local authority contracts) (Clause 30.8).
24. Fluctuations (Clause 37):
    Clause 38, 39 or 40 to apply.
    38.7 or 39.8 – Percentage addition.
    40.1.1.1 – Base month (Rule 3)
    Non-adjustable element (Rule 3), Part I/Part II of Section 2 to apply (Rules 10 and 30(i)).
25. Local authority fees.
26. Particulars of any other special conditions not normally anticipated.

SPECIFICATIONS

27. Temporary roads, etc.
28. Temporary buildings:
    (a) For Contractor.
    (b) For Architect and others.
    (c) For Clerk of Works.
    (d) Rates on foregoing.
29. Temporary telephones:
    (a) Particulars.
    (b) Cost of Employer's calls.
30. Temporary screens.
31. Temporary hoardings and gantries, and advertising rights on same.
32. Any special watching and lighting.
33. Provision for supplying of samples.
34. Provision for testing materials.
35. Provision for fuel for drying out building.
36. Provision for removing water below water table.
37. Particulars of other works expected to be in progress on or adjacent to the site.
38. Contingencies.

C. Demolitions and Alterations

1. General particulars.
2. Old materials to be retained by employer.
3. Old materials to be re-used.
4. Temporary screens etc.
5. Special shoring.
6. Particular items of covering and protecting existing fittings, adjoining property, etc.

D. Excavation and Earthwork

1. Site preparation.
2. Datum level.
3. Site levels and floor levels.
4. Nature of sub-soil.
5. Trial holes and existing services.
6. Level of sub-soil water and details of tide levels for river and sea.
7. Return fill in and ram.
8. Disposal of surplus soil.
9. Hardcore:
    (a) Type of material.
    (b) Maximum size of material.
    (c) Thickness of beds.
    (d) Weight of roller to be used.
    (e) Finishing for concrete.

E. Piling and Diaphragm Walling

1. Nature of ground and strata.
2. Type of piling giving full details.
3. Level that piling is to commence.
4. Precautions to be taken relating to adjoining properties.
5. If piling is carried out by a specialist then builder's work items:
    (a) Setting out.
    (b) Disposal of pile excavations.
    (c) Cutting off pile caps.
    (d) Bending reinforcement at top of piles.

F. Concrete Work

1. Cement.
2. Fine aggregate.
3. Coarse aggregate.
4. Tests.
5. Frost.
6. Proportions of mix and where to be laid.
7. Finish to wrought formwork and where to be used.
8. Reinforcement.
9. Expansion joints and construction joints.

10. Damp proofing.
11. Hardening and dust proofing.
12. Key for plaster.
13. Precast concrete.
14. Hollow tile and precast concrete slabs.
15. Prestressed concrete.
16. Pavement lights.
17. Column guards and masonry anchors.
18. Building paper.
19. Contractor designed construction giving full details.

G. Brickwork and Blockwork

1. Common bricks:
   (a) Below dpc.
   (b) Above dpc.
2. Facing bricks:
   (a) Below dpc.
   (b) Above dpc.
3. Special bricks.
4. Bonds of brickwork.
5. Mortar mixes.
6. Pointing:
   (a) Facing bricks.
   (b) Fair faced work internally.
7. Ties in cavity walls – type and spacing.
8. Damp-proof courses.
9. Partitions:
   (a) Type of block.
   (b) Mortar mix.
10. Decorative brickwork.
11. Air bricks.
12. Flue linings.
13. Throat units.
14. Chimney pots.
15. Soot doors.
16. Reinforcement.
17. Method of closing cavities:
    (a) at jambs.
    (b) at cills.

18. Vertical joints between brickwork and concrete.
19. Vertical joints between brickwork and partition blocks.
20. Horizontal joints between partitions and ceilings.
21. Glass blocks:
    (a) Type and size.
    (b) Mortar mix.
    (c) Mastic.
    (d) Reinforcement.
22. Gas flues.
23. Fireplaces (including fires, back boilers, stoves, surrounds and provision for gas ignition).
24. Slate and terrazzo cills and the like.

H. Underpinning

1. Description of structure to be underpinned giving:
   (a) Location.
   (b) Length on plan.
   (c) Depth of foundation below ground level.
2. Description of new work giving:
   (a) Depth below existing foundation.
   (b) Limit of length of each operation.
   (c) Details of materials to be used including mix of concrete or type of brickwork.
   (d) Thickness of new wall.

J. Rubble Walling

1. Particulars of stone or quarry.
2. Type of walling.
3. Details of coursing.
4. Finish.
5. Mortars:
   (a) Bedding.
   (b) Pointing

SPECIFICATIONS

K. Masonry

For each of the following:
(A) Natural stone (each type).
(B) Artificial and reconstructed stone. State as applicable:
1. Quarry.
2. Bed.
3. Mix.
4. Aggregate.
5. Colour.
6. Texture and finish.
7. Mortars:
    (a) Bedding.
    (b) Pointing.
8. Bonding to backing.
9. Treatment of back of stone.
10. Cramps.
11. Dowels.
12. Joggles.
13. Lengths of stones (sections as on drawings):
    (a) Copings.
    (b) Cills and string courses.
    (c) Mullions and jambs.
    (d) Columns and pillars.
14. Protection.
15. Sculpture and carving:
    (a) Where to be executed (on or off site).
    (b) Weight if executed off site.

L. Asphalt Work

For each of the following:
(A) Horizontal tanking.
(B) Vertical tanking.
(C) Roof covering.
(D) Pavings.
State:
1. British Standard.
2. Thickness (and heights of skirtings).
3. Number of layers.
4. Colour.
5. Finish.
6. Felt underlay.
7. Reinforcement.
8. Skirtings and upstands to roof.
9. Skirtings and upstands to pavings.

M. Roofing

1. For tiles, slates, etc., state:
    (a) Description of materials.
    (b) Size.
    (c) Gauge.
    (d) Method of fixing.
    (e) Size of battens.
    (f) Finish at eaves.
    (g) Finish at verges.
    (h) Finish at abutments.
    (i) Ridges, including ends.
    (j) Hips, including ends.
    (k) Valleys.
    (l) Special fittings, e.g. glass tiles and vent tiles.
    (m) Underfelt.
2. For metal and other corrugated coverings, state:
    (a) Description of material.
    (b) End and side laps.
    (c) Method of fixing.
    (d) Finish at eaves.
    (e) Finish at verges.
    (f) Finish at abutments.
    (g) Ridges, including ends.
    (h) Hips, including ends.
    (i) Valleys.
    (j) Expansion joints.
    (k) Special fittings, e.g. ventilators and roof lights.
3. Roof decking:
    (a) Description of material giving kind of decking, thickness, quality and method of fixing.

PRE-CONTRACT PRACTICE

    (b) Details of fixing blocks.
    (c) Details of all bearings, eaves, kerbs, flashings and nibs.
4. Felt roofing:
    (a) Type of felt.
    (b) Number of layers.
    (c) Weight of each layer.
    (d) Bonding.
    (e) Applied finish.
    (f) Lapping and jointing.
    (g) Falls.
    (h) Finish at eaves.
    (i) Finish at verges.
    (j) Finish at abutments.
    (k) Treatment of ridges and hips.
    (l) Treatment of valleys.
    (m) Treatment of outlets.
    (n) Working around pipes, balusters, etc.
5. Sheet metal roofing:
    (a) For each of the following:
        (i) Lead.
        (ii) Zinc.
        (iii) Copper.
        (iv) Aluminium.
State:
    (b) Weight or gauge.
    (c) Maximum size of sheets.
    (d) Methods of jointing and fixing.
    (e) Underlay.
    (f) Flashings.
    (g) Soakers.
    (h) Aprons.
    (i) Wedgings for flashings.
6. Thatching.
    (a) Description of material giving kind of thatch, thickness, and method of fixing.
    (b) Finish at abutments.
    (c) Finish at eaves.
    (d) Finish at verge.
    (e) Finish at hips.
    (f) Finish at ridge.
    (g) Finish at valleys.
    (h) Details of ornamental features.
    (i) Description of wire netting if required.
7. Sheet metal flashings and gutters giving full details.

N. Woodwork

Carcassing and First Fixings

1. Type and/or stress grading of timber.
2. Preservatives.
3. Strutting for floor joists.
4. Battens to edges of suspended ceilings.
5. Noggings for plasterboard and other linings.
6. Firrings.
7. Insulation in floors and roofs.
8. Snowboards and gang boarding.
9. Cistern casings.
10. Flag staffs.
11. Tie rods, straps, etc.
12. Bolts and other connectors.

Second Fixings

1. Types of timber.
    (a) Softwoods.
    (b) Hardwoods.
    (c) Ply, blockboard, chipboard, hardboard, etc.
2. For each of the following:
    (A) Boarded flooring.
    (B) Strip flooring.
State:
    (a) Timber.
    (b) Jointing.
    (c) Margins.
    (d) Finish.

SPECIFICATIONS

3. Plain and matchboarded linings.
4. Panelled linings.
5. Particulars of the following where not shown on detail drawings or schedules:
   (a) Doors and frames.
   (b) Windows and frames.
   (c) Borrowed lights.
   (d) Lantern lights.
   (e) Hatches.
   (f) Skirtings.
   (g) Cornices.
   (h) Friezes and dados.
   (i) Sub-frames.
   (j) Cupboard units.
   (k) Pelmets.
   (l) Cloak rails.
   (m) Pipe casings and access doors.
   (n) Backboards.
   (o) Picture rails.
   (p) Architraves and cover fillets.
   (q) Window boards.
   (r) Staircases.
   (s) Shelving.
   (t) Trap doors.
   (u) Other fittings.
6. Information in the following where not shown on detail drawings or schedules:
   (a) Ironmongery for doors.
   (b) Ironmongery for windows.
   (c) Ironmongery for fittings.
   (d) Fixing cramps for joinery.
   (e) Dowels for joinery.
   (f) Water bars.
   (g) Shelf brackets.
   (h) Handrail brackets.
   (i) Hat and coat hooks.
   (j) Mats.
   (k) Lettering and numerals.
   (l) Curtain tracks.
   (m) Special ironmongery.

P. Structural Steelwork

1. Structural steel:
   (a) Shop finish.
   (b) Site treatment.
   (c) Holding-down bolts.

Q. Metalwork

1. Metal windows, doors and rooflights:
   (a) Material.
   (b) Section.
   (c) Fixing lugs.
   (d) Shop finish.
   (e) Glazing beads.
   (f) Ironmongery.
   (g) Gearing.
   (h) Curtain track brackets.
   (i) Bedding and pointing.
   (j) Sub-frames and cills.
2. Curtain walling.
3. Particulars of the following when not shown on detail drawings:
   (a) Railings and balustrades.
   (b) Staircases and balconies.
   (c) Cat ladders and gangways.
   (d) Fire doors.
   (e) Shutters.
   (f) Collapsible gates.
   (g) Other doors and gates.
   (h) Ventilators and grilles.
   (i) Shop fittings.
   (j) Steel partitions.
   (k) Blinds.
   (l) Refuse hoppers and fittings.
   (m) Mat frames.
   (n) Trim.
4. Particular items of covering and protecting finished work.

PRE-CONTRACT PRACTICE

R. Plumbing and Mechanical Engineering Installations

1. For each of the following:
   (A) Rainwater gutters.
   (B) Rainwater pipes.
   State:
   (a) Material.
   (b) Size.
   (c) Type.
   (d) Method of fixing.
   (e) Method of jointing.
   (f) Connection of pipes to drains.
2. Roof outlets.
3. Rainwater heads.
4. Gratings.
5. Connection to Company's main.
6. Pipes
   For each of the following:
   (a) External water main.
   (b) Rising main.
   (c) Cold water services.
   (d) Hot water services.
   (e) Overflows.
   (f) Wastes.
   (g) Soil pipes.
   (h) Vent pipes.
   (i) Expansion pipes.
   State:
   (A) Material.
   (B) Method of jointing.
   (C) Method of fixing.
   (D) Type of fittings.
7. Stop valves.
8. Bib valves.
9. Drain off cocks.
10. Safety valves.
11. Ball valves.
12. Meters.
13. Sanitary fittings:
    Give particulars or state name of supplier and reference to quotation and/or schedule.
14. Traps:
    (a) Material.
    (b) Type.
15. Cisterns and tanks – for (A) cold and (B) hot water state:
    (a) Type and quality.
    (b) Capacity.
    (c) Size.
16. Lagging and other insulation:
    (a) Pipes.
    (b) Cisterns.
17. Ductwork. Details for heating and ventilation systems will usually come from a specialist consultant and, if a bill of quantities is to be produced, full details will have to be provided.

S. Electrical Installation and Specialist Services

Notes

1. It is *not* anticipated that the architect will normally prepare information from which the quantity surveyor can directly produce a bill of quantities for these specialist services, but rather that this will come from a consultant or alternatively that a PC sum will be included in the bills based on tenders from specialist sub-contractors.

2. It may well be that the architect will find it convenient to prepare some information for the consultants in the form of specification notes, either in the nature of a performance specification or, on the less technical items, particulars of specific requirements, for example, type, material and colour of electric switches, but the possibilities and alternatives are too numerous to make a tabulated proforma workable.

SPECIFICATIONS

3. The following information is intended only as a guide to the quantity surveyor to the sources from which he is to expect his information.

For each of the following Specialist Services:
1. Electric main.
2. Electrical services.
3. Gas main.
4. Gas services.
5. Heating.
6. Hot water.
7. Ventilation.
8. Sprinklers and dry risers.
9. Lifts, escalators and conveyors.
10. Mechanical plant.
11. Kitchen equipment.
12. Lightning conductors.
13. Telephones.
14. Bell and call systems.
15. Radio and television system.
16. Earthing systems.

Give name and address of consultant if any and state method by which each service will be dealt with in bill, e.g.:
    (a) Measured in bills of quantities.
    (b) PC from sub-contract bills of quantities.
    (c) PC from specification and drawings – or by selection.
    (d) PC as (c) but design developed by sub-contractor.
    (e) PC from artist or craftsman.
    (f) PC from local authority or public utility.

Work in Connection with Services

Note: Much of this work, particularly trenches, holes, chases and the like, will be self explanatory from the drawings and particulars of the services involved prepared by the architect or his consultant. The following items, however, may require special mention:

1. Stop cock and valve pits:
    (a) Sizes.
    (b) Materials.
    (c) Covers.
2. Bases for pumps, motors, boilers and other equipment:
    (a) Sizes.
    (b) Materials.
    (c) Mortices.
    (d) Special finishes.
    (e) Ash pits.
3. For each of the following:
    (A) Ducts in floors.
    (B) Ducts in walls.
State:
    (a) Sizes.
    (b) Pipe bearers.
    (c) Filling.
    (d) Covers.
    (e) Access traps.
4. Pipe casings:
    (a) Sizes.
    (b) Materials.
    (c) Access panels.
5. Tank bearers.
6. Tank casings.
7. Pipe sleeves.
8. Hangers and brackets to be fixed by general contractor.
9. Soot doors.
10. Draught stabilisers.
11. Backboards for sanitary fittings and equipment.
12. Floor channels:
    (a) Type.
    (b) Size.
    (c) Outlets.
    (d) Covers.

PRE-CONTRACT PRACTICE

13. Bearers or frames for electric switch gear.
14. Cupboards for electric switch gear and meters.
15. Fuel boards and baffles.
16. Fuel for testing installation.
17. Fittings when not included with installation:
    (a) Lighting fittings.
    (b) Cookers.
    (c) Refrigerators.
    (d) Wash boilers.
    (e) Water heaters.
    (f) Gas fires.
    (g) Other unit heaters.
    (h) Pumps.
    (i) Fuel bins.
    (j) Water softeners.
    (k) Machinery.
    (l) Kitchen equipment.
    (m) Fire fighting equipment.
18. Paint on pipes (if different from that specified in 'Painting and Decorating'):
    (a) Cold water pipes.
    (b) Waste and soil pipes.
    (c) Hot water pipes.
    (d) Heating pipes.
    (e) Gas pipes.
    (f) Electric conduit.
    (g) Other service pipes.
19. Further works in connection with specialist services.

T. Floor, Wall and Ceiling Finishings

1. For each type of paving or floor covering state:
   (a) Description.
   (b) Mix.
   (c) British Standard.
   (d) Thickness and size of units.
   (e) Colour.
   (f) Screed or floated bed.
   (g) Bedding and jointing.
   (h) Adhesive.
   (i) Hardening, dust proofing and sealing.
   (j) Polishing.
   (k) Non-slip finish.
   (l) Expansion joints and separation strips.
   (m) Skirtings to pavings.
2. For each of the following:
   (a) Block flooring.
   (b) Parquet flooring.
   State:
   (A) Timber.
   (B) Jointing.
   (C) Margins.
   (D) Finish.
3. Metal lathing:
   (a) Gauge and mesh.
   (b) Surface finish.
   (c) Method of fixing.
4. Plasterboard:
   (a) Type of board.
   (b) Method of fixing.
   (c) Jointing and scrimming.
   (d) Plaster finish.
5. Acoustic tiles.
6. Bonding agents.
7. Ceiling and wall plaster:
   (a) Number of coats.
   (b) Type and mix of plaster for each coat.
   (c) Finish to salient angles.
8. Coves and cornices, including bracketing.
9. Fibrous plaster.
10. Asbestos cement wall and ceiling linings, type and fixing.
11. Wall boards, type, thickness and fixing.
12. Fabric and similar wall coverings, type and fixing.
13. Staircase and step finishes:
    (a) Treads.
    (b) Risers.

SPECIFICATIONS

    (c) Strings.
    (d) Skirtings.
    (e) Nosings.
    (f) Landings.
14. Renderings:
    (a) Types.
    (b) Number of coats or thickness.
    (c) Mix for each coat.
    (d) Method of application.
    (e) Finish.
15. Cement and sand screeds:
    (a) Mix.
    (b) Thickness and falls.
    (c) Waterproofing.
16. Lightweight screeds:
    (a) Aggregate.
    (b) Mix.
    (c) Thickness and falls.
    (d) Toppings.
17. Cement glaze.
18. Terrazzo:
    (a) Aggregate and matrix.
    (b) Thickness.
    (c) Mix.
    (d) Finish.
    (e) Separation strips.
    (f) External angles.
    (g) Internal angles.
19. Wall tilings:
    (a) Type and quality.
    (b) Colour.
    (c) Thickness and size.
    (d) External angles.
    (e) Internal angles.
    (f) Bands.
    (g) Method and type of bedding, jointing and pointing.
20. Mosaic:
    (a) Type.
    (b) Colour.
    (c) Thickness and size.
    (d) External angles.
    (e) Method and type of bedding, jointing and pointing.

U. Glazing

1. For each sort of glazing state:
    (a) Type of glass.
    (b) Weight or thickness.
    (c) Quality.
    (d) Method of glazing.
2. Patent glazing.
3. Roof lights.
4. Mirrors and other glazing accessories.
5. Particular items of covering and protecting finished work.

V. Painting and Decorating

1. For each of the following:
    (a) Distemper.
    (b) Cement based paint.
    (c) Emulsion paint.
    (d) Paint on metal externally.
    (e) Paint on metal internally.
    (f) Paint on galvanised metal externally.
    (g) Paint on galvanised metal internally.
    (h) Paint on wood externally.
    (i) Paint on wood internally.
    (j) Paint on other surfaces.
State:
    (A) Name of manufacturer.
    (B) Preparation of surfaces.
    (C) Primer.
    (D) Type of paint and number of undercoats.
    (E) Type of paint and number of finishing coats.
Particulars of the following if required:
2. Work in parti-colours.
3. Staining.
4. Wax polish.

PRE-CONTRACT PRACTICE

5. French polish.
6. Lettering.
7. Other types of finish.
8. Paper-hanging:
   (a) Type of paper and PC.
   (b) Method of hanging.
   (c) Preparation.
9. Particular items of covering and protecting finished work.

W. Drainage

1. For each of the following:
   (A) Rainwater drains.
   (B) Soil drains.
   (C) Drains for special effluents.
   State:
   (a) Type and grade of pipe.
   (b) Jointing.
   (c) Particulars of bed or surround.
2. Land drains.
3. Gullies and traps:
   (a) Types.
   (b) Gratings.
   (c) Sealing plates.
   (d) Raising pieces.
   (e) Inlets.
   (f) Kerbs.
4. Cleaning eyes.
5. Connections to sewer:
   (a) To be done by.
   (b) Approximate cost.
6. Manholes.
7. Fresh air inlets.
8. Interceptors for petrol and other special effluents.
9. Septic tanks.
10. Cesspools.
11. Soakaways.
12. Sewage disposal plants.

X. External Works and Fencing

For each of the following:
(A) Roads.
(B) Paths.
(C) Other pavings.
State:
1. Excavation or filling.
2. Base – material and thickness.
3. Finish – material and thickness.
4. Weight of roller for consolidation.
5. Expansion joints.
6. Kerbs, channels and edgings, spur stones and bollards, and roadmarking.
7. For each of the following:
   (A) Boundary walls.
   (B) Fences.
   State:
   (a) Type.
   (b) Height.
   (c) Foundations or bases.
   (d) Piers or posts.
   (e) Filling between piers or posts.
8. Gates.
9. Crossovers.
10. External signs.
11. Road lighting:
    (a) Standards.
    (b) Fittings.
    (c) Bases for standards.
    (d) Trenches for cables.
    (e) Other work in connection.
12. Laying services mains:
    (a) Water.
    (b) Gas.
    (c) Electricity.
13. Other external work in connection with services.
14. Garden works:
    (a) Preparation.
    (b) Seeding and turfing.
    (c) Hedging and ditching.

SPECIFICATIONS

(d) Planting of trees, shrubs and flowers.

(e) Protection of trees, shrubs and hedges.

Application of these specification headings in practice is shown in Example J.

Probably the best procedure is for the architect to prepare the notes as he develops his design and then for the quantity surveyor to go through them with him, making his own copy and dealing with any queries or points of clarification that may arise.

When completing the notes against the headings, matters already adequately dealt with on drawings or schedules should not be repeated, but a simple cross reference given. The notes should be brief and to the point, and long descriptions can often be avoided by the inclusion of sketches.

Wherever applicable, reference should be made to British Standard Specifications and Codes of Practice. Such references should give not only the Standard or Code number, but should also be quite specific as to which of any alternatives within the Standard or Code are required.

Similarly, if reference is made to the National Building Specification great care must be taken to ensure that all the appropriate clauses relevant to the item in question are included and that spaces left blank in the NBS descriptions are filled in. It should also be noted that the NBS refers frequently to British Standards and here again specific reference to alternatives within the British Standard Specification must be made.

Word Processing

Word processors can be of great assistance in specification writing. The ease with which insertions, amendments, deletions and the re-ordering of clauses can be effected, without multiple re-typing, makes the process much quicker and less labour intensive. The document can be built up gradually as the dialogue between the architect and quantity surveyor develops, or as the detailed design is refined, and the final docment can be produced very quickly after the last adjustments have been made.

Quite sophisticated word processing packages can be obtained for even quite modest microcomputers and software packages for standard specification documents, such as the NBS, are readily available.

PRE-CONTRACT PRACTICE
EXAMPLE J EXAMPLES OF SPECIFICATION NOTES

G. BRICKWORK AND BLOCKWORK

1	Common bricks :	
	(a) Below dpc	Second Quality Stocks.
	(b) Above dpc	Flettons.
2	Facing bricks :	
	(a) Below dpc	Aqua Brick Co. Golden Glory No.2.
	(b) Above dpc	
3	Special bricks	38 mm Slips to face of beam on west elevation. (Aqua Brick Co.)
4	Brickwork bonds	Stretcher to hollow walls. English elsewhere.
5	Mortar mixes	Cement and Sand (1:3) below d.p.c. Cement, Lime, sand (1:2:9:) above d.p.c.
6	Pointing:	
	(a) Facing bricks	Weathered Struck as work proceeds.
	(b) Fair faced work internally	Flush.
7	Wall ties and spacing	B.S.1243. Type a Butterfly 150 mm Long. Spaced 1 m. horizontally and 500 mm vertically staggered.
8	DPC's	"Aquaproofo" bit.felt grade "P" 100 mm Laps.
9	Partitions:	
	(a) Type	Clinker conc. as B.S.2028 Type C.
	(b) Mortar	Cement and Sand (1:4)
10	Decorative brickwork	Blue Staffordshire projecting headers (drwg. 456/4)
11	Air bricks	225 x 150 mm. T.C. square hole.
12	Flue linings	None.
13	Throat units	None.
14	Chimney pots	None.
15	Soot doors	None.
16	Reinforcement	None.
17	Closing of cavities:	
	(a) At jambs	Return inner skin of Bwk. 112.5 mm wide vertical d.p.c.
	(b) At cills	Closed by Art.Stone Cills (See Masonry)

SPECIFICATIONS

.8	Vertical joints : brickwork to concrete	Galvanised hoop Iron Cramps cast into Conc. at 500 mm c/c and built into joints.
19	Vertical joints : Brickwork to blockwork	Bond in alternate courses of partitions.
20	Horizontal joints : partitions to ceilings	Pinned up with slates.
21	Glass blocks : (a) Type and size (b) Mortar (c) Mastic (d) Reinforcement	None.
22	Gas-flues	None.
23	Fireplaces, back boilers, stoves, surrounds, etc.	None.
24	Slate and terrazzo cills, etc.	None.

H. UNDERPINNING

1	Structures to be underpinned (a) Location (b) Length on plan (c) Depth of foundation	None.
2	Description of new work : (a) Depth below existing foundation (b) Limit of length of each operation (c) Materials to be used (d) Thickness of new wall	None.

PRE-CONTRACT PRACTICE

L. ASPHALT WORK

		(A) Horizontal tanking	(B) Vertical tanking	(C) Roof covering	(D) Pavings
1	B.S.	1418	1418	1162	1410
2	Thickness (and heights of skirtings)	30 mm	20 mm	20 mm	25 mm
3	Number of layers	3	3	2	1
4	Colour	—	—	—	Black
5	Finish	Subsequently covered		13 mm Limestone Chippings	Sanded
6	Felt Underlay	—	—	B.S.747 Class 4A	—
7	Reinforcement	—	—	—	—
8	Skirtings and Upstands to Roof	—	—	20 mm 2 coat. 150 mm high.	—
9	Skirtings and Upstands to Pavings	—	—	—	20 mm 1 coat 100 mm high.

Chapter 6

BILLS OF QUANTITIES

Bills of quantities are contract documents and as such have a specific role to fulfil under the standard form of contract. It is therefore important that they are prepared in accordance with the conditions of the contract, contain certain basic information, and are presented in a recognisable format which will facilitate their use.

However, it has long been recognised that, in preparing this basic information, the quantity surveyor processes a great deal of detailed information, much of which could be made available to the contractor. This would be of use, not only in tendering, but also in contract planning and administration. It is therefore always worthwhile, in the early stages of any job, spending some time considering the role the bills are required to play and what additional information could be of use to the particular parties involved.

The Role of Bills of Quantities

Although primarily designed as tendering documents, bills of quantities have an important contractual role to play in the pricing of variations. These variations, with the original contract sum, will form part of the final account. Additionally, bills of quantities are usually used in the computation of valuations for interim certificates.

There are, however, a number of additional roles that the bills can play. Of these the two most important are the locational identification of the work and the formation of a basis for cost planning, both of which will be discussed in more detail below.

Basic Information

The basic information contained in bills of quantities falls into three categories, namely:

- preliminaries,
- preambles or descriptions of materials and workmanship,
- measured works,

and within these three sections should be contained a complete description of the works and the conditions under which those works

PRE-CONTRACT PRACTICE

are to be carried out. In this way the bills of quantities become a co-ordinated whole and a complete financial representation of the works, in support of the contractor's tender.

The PRELIMINARIES should contain a definition of the scope of the works and details of the proposed form of contract, indicating the employer's intention in relation to all the options within the contract, as well as the details required to complete the Appendix. All proposed amendments to the form of contract should be set out in full, although it is recommended that the standard forms are used without amendment, whenever possible. The preliminaries should contain a detailed description of the administrative mechanisms that will be necessary to implement the conditions of contract and any special conditions as to the way in which the works are to be carried out. The preliminaries should also contain a list of the drawings upon which the documents are based and any special instructions in the method of pricing and presentation of the contractor's tender.

The PREAMBLES or Descriptions of Materials and Workmanship are intended as a concise definition of all the materials required and of the standard or quality of workmanship to be employed in working or assembling them. This enables the measured work section to be kept reasonably brief and to be easily priced by contractor's estimators. These premables must convey, either directly or by reference, the architect's and engineer's requirements as defined in their respective specifications.

The MEASURED WORKS section or sections, which now often include fully measured mechanical and electrical services as well as building works, should form a detailed description of the works, presented in accordance with the rules of measurement laid down under the contract (usually the Standard Method of Measurement of Building Works – SMM). This enables the contractor's estimator to price each individual item of work according to recognised conventions. The measured works will usually also contain prime cost (PC) and provisional sums. It is important to recognise the difference between these two facilities and to appreciate the procedures that are laid down for their use in the standard forms of contract (Chapter 7 describes in detail the procedures for sub-contractors and suppliers). For convenience a brief definition of each follows.

- *Prime cost sums* are used for works to be carried out by nominated sub-contractors and statutory authorities or for goods to be supplied by nominated suppliers, for which the sub-contractor or supplier will usually have been selected in the pre-contract stage and for which estimates or tenders should have been obtained. The JCT Standard Form of Nominated Sub-Contract Tender and Agreement (Tender NSC/1) is designed for this purpose – see Chapter 7. Provision will be made in the bills for the contractor to add his profit to these sums and also, in the case of sums for sub-contract work, items will be included for general attendance and such items of special attendance as have been identified in the JCT standard form NSC/1 or in discussions with the particular specialists at the pre-contract stage. Sums for nominated suppliers will generate items for the 'fixing only' of the goods concerned.

- *Provisional sums* are used for work which cannot be fully defined, or for costs which are unknown at the time the bills are prepared. The architect must issue instructions, in accordance with the contract conditions, regarding the expenditure of these sums, which may cover measured builders' work, further nominated sub-contractors' works or nominated suppliers' goods. They must therefore be sufficient to cover the full cost of the work involved which, in the case of a nomination, includes the profit and general and special attendance items, and also, where appropriate, additional preliminaries.

Additional Information

The desire to make bills of quantities more useful usually leads to the provision of additional information which involves the retrieval or classification of the detailed information identified in the measurement process. The specific use usually dictates the type and format of the extra information provided.

The increased complexity of projects has led to a need for some locational identification of the items of work described in the bills and this has encouraged the development of the locational analysis of bills, operational bills and annotated bills.

The increased use of cost planning techniques has led to the development of elemental bills and the elemental analysis of items included in traditional bills.

A brief description of these different documents follows and examples of the more common ones are set out at the end of this chapter. It is important to note that all the different formats are produced from the same basic measurement process and that, as long as the requirements are known before measurement starts, there is no reason why the information should not be presented in any of the alternative formats available. Indeed, as long as the correct coding is done during the original measurement, there is no reason why the quantities should not be produced in one format and later be re-sorted for an alternative presentation. Computer processing of quantities can make this particularly easy.

Bill Formats

Traditional bills of quantities are presented under trade or work section headings with the quantities for each item being presented in a single total irrespective of location (Example K). Trade totals are collected in a summary to ascertain the total value of the works.

Additional information can be added to this presentation by providing extra columns in the bill for a location or elemental breakdown of the total quantities (Example L). With this extra breakdown the location or elemental cost totals can quickly be calculated for cost planning purposes.

If the cost planning function is of prime importance, the bills can be presented in elemental format (Example M). In this case the work is broken down into standard elements, normally those used by the Building Cost Information Service (BCIS) of the Royal Institution of Chartered Surveyors. When broken down in this way, the total quantities of any particular trade may not be immediately available, being split between elements, and this can be inconvenient for contractors' estimators, especially when obtaining sub-contract prices. However, with modern computer documentation it is possible to overcome this difficulty by having the information appropriately coded and produced in two, or even more, formats – a trade bill for tendering, an elemental bill for cost planning and an operational bill for production control or contract administration.

Occasionally there is a need for an even more specialised format for the bills, for instance, where facilities for the analysis and control of labour resources are required, as in direct labour contracts. In these cases an operational bill may be appropriate. The operational bill can be presented broadly in either traditional trade or elemental format, but with the difference that the work items are finally broken down into labour operations or activities. These operations or activities might best be likened to the breakdown that a contractor would produce for bonus payments to his work force.

If greater detail of the location of the billed items is required, the annotated bill can be used. Annotated bills have been in use for many years and have appeared in several forms. The annotations can be provided as a separate document, they can be bound into the back of the bill or they can be reproduced directly opposite each item in the document as shown in Example N. Annotations can be added to any of the formats previously described and can include reference to detailed drawings or schedules as well as amplified descriptions or simple locational references.

Further Developments

As in other fields of communication, new techniques and procedures are always being tried and bills of quantities can, and will, change according to the needs of the parties to the project. It is up to the building team to decide how best the flexibility of the system can be used to the benefit of all concerned, not forgetting that efficient documentation should lead to efficient working and hopefully to some cost advantage to the client.

PRE-CONTRACT PRACTICE

EXAMPLE K TRADITIONAL TRADE OR WORK SECTION FORMAT

Note: Extracts from three separate sections are set out in order to allow comparison with other formats. The order of the sections, and the items within them, generally follows the layout of the SMM.

Item No.	BRICKWORK and BLOCKWORK			£
	BLOCKWORK			
	Lightweight Blocks in Gauged Mortar			
a	100 mm wall	427	m²	
b	100 mm skin of hollow wall	1236	m²	
c	100 mm wall finished fair and flush pointed one side	78	m²	
d	Fair return 100 mm wide	38	m	
	etc.			
	FLOOR, WALL AND CEILING FINISHES			
	IN SITU FINISHING – INTERNALLY			
	Two-coat Lightweight Gypsum Plaster on the Following			
h	Brick walls	860	m²	
j	Block walls	2168	m²	
k	Concrete ceiling	1435	m²	
l	Side of isolated concrete column not exceeding 300 mm wide	22	m²	
m	Slightly rounded external angle	511	m	
	etc.			
	PAINTING AND DECORATING			
	NEW WORK INTERNALLY			
	Preparation, One Mist Coat and Two Full Coats Emulsion Paint on the Following			
s	Plastered wall	3050	m²	
t	Plastered ceiling	1435	m²	
	Carried to Collection			£

BILLS OF QUANTITIES

EXAMPLE L TRADE OR WORK SECTION FORMAT WITH ANALYSIS

Note: Extracts from two formats are set out. In the first example, the breakdown into categories A, B and C etc. could be used to identify different house types, buildings or phases. In the second example, the breakdown is into the BCIS standard list of elements and their references are used (see Chapter 2).

Item	BRICKWORK and BLOCKWORK					£
	BLOCKWORK					
	Lightweight Blocks in Gauged Mortar					
a	100 mm wall	A B C	340⎫ 67⎬ 20⎭	427	m²	
b	100 mm skin of hollow wall	A B C	976⎫ 192⎬ 68⎭	1236	m²	
c	100 mm wall finished fair and flush pointed on one side	A B C	45⎫ –⎬ 33⎭	78	m²	
d	Fair return 100 mm wide	A B C	21⎫ –⎬ 17⎭	38	m	
	BLOCKWORK					
	Lightweight Blocks in Gauged Mortar					
A	100 mm wall	2G	427	427	m²	
B	100 mm skin of hollow wall	2E	1236	1236	m²	
C	100 mm wall finished fair and flush pointed one side	2D 2G	30⎫ 48⎭	78	m²	
D	Fair return 100 mm wide	2D 2G	32⎫ 6⎭	38	m	
		Carried to Collection			£	

PRE-CONTRACT PRACTICE

"... *presented in a recognisable format* ..."

BILLS OF QUANTITIES

EXAMPLE M ELEMENTAL FORMAT

Note: Extracts from two elements are set out, the BCIS element references being quoted. The trade or work section headings appear, as appropriate, as subsidiary side headings, under the main elemental sections.

Item				£	p
	Internal walls and partitions (Element 2.G)				
	BRICKWORK and BLOCKWORK				
	BLOCKWORK				
	Lightweight Blocks in Gauged Mortar				
a	100 mm wall	427	m²		
b	100 mm wall finished fair and flush pointed one side	48	m²		
c	Fair return 100 mm wide	6	m		
	etc.				
	Wall finishes (Element 3.A)				
	FLOOR, WALL AND CEILING FINISHES				
	IN SITU FINISHES – INTERNALLY				
	Two Coat Lightweight Gypsum Plaster on the Following				
g	Brick walls	860	m²		
h	Block walls	2168	m²		
j	Side of isolated concrete column not exceeding 300 mm wide	22	m²		
k	Slightly rounded external angle	511	m		
	etc.				
	PAINTING AND DECORATING				
	NEW WORK – INTERNALLY				
	Preparation, One Mist Coat and Two Full Coats Emulsion Paint on the Following				
r	Plastered walls	3050	m²		
s	Fair face block walls	49	m²		
	Carried to Collection			£	

75

PRE-CONTRACT PRACTICE

EXAMPLE N ANNOTATED BILL

Note: Annotations are best set out on the facing page opposite the items they amplify.

Annotations

Item

a	Non-load bearing partitions first floor (drwg 456/78).
b	Internal skin of hollow walls generally.
c	Stores ground and first floor and staircase enclosure walls.
d	See item c.

h	Load bearing partitions ground floor.
j	Internal skin of hollow walls, first floor partitions and external face of store walls.
k	Soffit of first floor.
l	Columns of hall.
m	Generally except where angle beads are used.

s	Refer to colour schedule.
t	Generally.

BILLS OF QUANTITIES

Item	BRICKWORK and BLOCKWORK			£	p
	BLOCKWORK				
	Lightweight Blocks in Gauged Mortar				
a	100 mm wall	427	m²		
b	100 mm skin of hollow wall	1236	m²		
c	100 mm wall finished fair and flush pointed one side	78	m²		
d	Fair return 100 mm wide	38	m		
	etc.				
	FLOOR, WALL AND CEILING FINISHES				
	IN SITU FINISHINGS — INTERNALLY				
	Two Coat Lightweight Gypsum Plaster on the Following				
h	Brick walls	860	m²		
j	Block walls	2168	m²		
k	Concrete ceiling	1435	m²		
l	Side of isolated concrete column not exceeding 300 mm wide	22	m²		
m	Slightly rounded external angle	511	m		
	etc.				
	PAINTING AND DECORATING				
	NEW WORK INTERNALLY				
	Preparation, One Mist Coat and Two Full Coats Emulsion Paint on the Following				
s	Plastered wall	3050	m²		
t	Plastered ceiling	1435	m²		
	Carried to Collection			£	

Chapter 7

SUB-CONTRACTORS AND SUPPLIERS

In the early editions of this book we expressed ourselves strongly in favour of reducing to the minimum the number of sub-contractors and suppliers nominated by the architect, advocating in effect that all, or at any rate nearly all, the work should be left in the hands of the general contractor who would decide from whom he would buy his materials and to whom he would sublet those parts of the work he did not carry out with his own labour.

The building process has changed over the years and our views have changed with it in successive editions of *Pre-Contract Practice*.

Nominated Sub-Contractors

Construction has become more complex and early decisions have to be made on various elements of the building before the architect can proceed with the detailed design. Steel and concrete structural frames, for example, must be worked out in considerable detail before the architect can design the cladding of them; and the cladding itself may be a proprietary system. Mechanical and electrical services, which are frequently extremely complex and often account for 30 per cent of the cost of the building, and sometimes 50 per cent or more, make considerable demands on space for plant rooms, ducting and the like, and the architect must know just what is required before he can finalise the storey heights and floor lay-outs. Thus it is essential for decisions on a number of fundamental elements of the building to be made very early on in the design, and this may make it necessary to select the principal specialist sub-contractors at the beginning of the design process. Sub-contractors involved in this way may, in fact, take on some of the detailed design themselves, as frequently happens in the case of structural steelwork, mechanical installations and so on, and invariably happens in the case of certain installations such as lifts.

This early involvement puts these sub-contractors into a special relationship with the employer and in due course creates the situation in which contractor and sub-contractor are required to enter into a contract with each other, though neither has had any say in the matter.

Standard forms of contract have in the past made special provisions for sub-contractors nominated in this way, but those provisions have

SUB-CONTRACTORS AND SUPPLIERS

had many shortcomings, both in the contractual relationship and in the responsibility for the design element. In addition the system has been abused by using nominations by way of prime cost sums in the bills of quantities to cover sections of the work which have not been properly designed in the pre-contract stage. Contractors have for a long time been justified in regarding a bill of quantities containing a large number of prime cost sums as a portent of inefficient and uneconomic building.

Whilst it is common practice for contractors to sub-let substantial parts of the works to sub-contractors of their own choice, they see nominated sub-contractors as being imposed upon them and tend to try to shift responsibility for them on to the architect. Conversely nominated sub-contractors have seen themselves in a special relationship with the architect and with a special status conferred upon them in the contract conditions.

Various attempts have been made over the years to clarify the position and to stop abuses of the procedure, but the arrangements have become increasingly unsatisfactory and disputes and litigation involving nominated sub-contractors have been common. The introduction of the Employer/Sub-Contractor Agreement went a long way to ironing out some of the problems, particularly in relation to design responsibility, delay, and direct payment by the employer to the sub-contractor in the event of the main contractor's default, but an extensive overhaul of the procedure has clearly been needed for a long time.

This overhaul came first in the 1980 edition of the Standard Form of Contract (JCT 80) and the principles and procedures involved are examined in some detail below. More recently, the Intermediate Form of Contract (IFC 84) has introduced a different concept, that of named sub-contractors.

It must be emphasised that these two systems are not alternative procedures to be selected at will in the course of a project. They are procedures specific to the particular forms of contract and must be operated as appropriate depending on the form of contract chosen.

In this book, though not dealing with the contractual provisions themselves in depth, we concentrate on JCT 80. This form imposes procedures and disciplines on architects and quantity surveyors in relation to nominations, even in the pre-contract stage, the object being to sort out at an early stage, and certainly before the contractor and sub-contractor are instructed to enter into contract, all those matters which could, if not defined in detail before the price is agreed, lead to dispute and possibly to delay at a time which may be critical later. It is necessary therefore to touch upon them and to summarise the matters which have now become mandatory in pre-contract practice.

JCT 80 identifies four types of arrangement under which work may be sub-let to a sub-contractor. Two of these relate to 'Domestic Sub-

PRE-CONTRACT PRACTICE

Contractors' and two to 'Nominated Sub-Contractors'.

A domestic sub-contractor may be:

- A sub-contractor to whom the main contractor sub-lets part of the work entirely at his, the main contractor's, discretion subject only to the consent of the architect.
- A sub-contractor selected by the main contractor at his sole discretion from a list of not less than three firms named in the bills of quantities.

A nominated sub-contractor is defined as a sub-contractor whose selection is reserved to the architect. He may be nominated by way of a prime cost sum or by being named in the bills of quantities or in an instruction regarding expenditure of a provisional sum or, subject to certain qualifications, in a variation order.

The architect must make his nomination in one of the two ways laid down in the contract, namely by:

- The basic method.
- The alternative method.

The Basic Method of Nomination

The basic method involves the full documentation which accompanies the Standard Form, namely:

- JCT Standard Form of Nominated Sub-Contract Tender and Agreement (Tender NSC/1).
- JCT Standard Form of Employer/Nominated Sub-Contractor Agreement (Agreement NSC/2).
- JCT Standard Form of Nomination of a Sub-Contractor where Tender NSC/1 has been used (Nomination NSC/3).
- JCT Standard Form of Sub-Contract for Sub-Contractors who have Tendered on Tender NSC/1 and executed Agreement NSC/2 and been Nominated by Nomination NSC/3 (Sub-Contract NSC/4).

Once the decision to nominate using the basic method has been taken the procedure laid down in the contract must be followed and it is obviously sensible to adopt it when the first enquiry is made to the prospective sub-contractor. There is no point in obtaining a preliminary estimate from the sub-contractor and later having to go back to him using the proper documentation and probably finding that the estimate has risen substantially because the earlier one had not allowed for the various conditions and obligations which will be required by the sub-contract. Even for the initial enquiry, therefore, the Tender NSC/1 and the Agreement NSC/2 should be used as these documents initiate a

SUB-CONTRACTORS AND SUPPLIERS

dialogue which, by stages, will resolve all the necessary details in a logical order – thus ensuring a fair price for a properly defined scope of work. It is not practical to achieve this level of co-ordination in one enquiry or submission at the preliminary enquiry stage.

Tender NSC/1 contains the Tender itself followed by Schedule 1, into which are inserted the particulars of the main contract and the sub-contract; two Appendices dealing with fluctuations; and Schedule 2 into which are inserted particular conditions relating to the sub-contract.

The procedure for making a nomination under the basic method is summarised below. Only the earlier stages of the procedure will occur during the pre-contract period, but the later stages are also listed as it is necessary to understand the whole sequence of events up to the actual nomination.

(1) The architect prepares Tender NSC/1, inserting in Schedule 1 the particulars of the main contract.
(2) At the same time the architect prepares the Agreement NSC/2.
(3) The architect sends the original and two copies of NSC/1 and the original NSC/2 to the proposed sub-contractor.
(4) The proposed sub-contractor completes Tender NSC/1 and the two copies and signs on page 1.
(5) The proposed sub-contractor executes Agreement NSC/2 under hand or under seal as instructed by the architect.
(6) The proposed sub-contractor returns to the architect the original documents and the copies of NSC/1.
(7) The architect signs the Tender NSC/1 and the two copies on page 1 as 'approved' on behalf of the employer.
(8) The employer executes Agreement NSC/2, again either under hand or under seal, retaining the original and sending the architect a certified true copy.
(9) The architect sends to the proposed sub-contractor the certified copy of the Agreement.

The procedure will not go beyond this point during the pre-contract stage, but subsequently will continue as follows after the main contract has been placed.

(10) The architect sends the main contractor a preliminary notice of nomination, accompanied by the original Tender NSC/1 and its two copies as then completed, together with a copy of the Agreement NSC/2 for the main contractor's information.
(11) The main contractor checks that the main details in Schedule 1 of Tender NSC/1 are correct.
(12) The main contractor completes Schedule 2 insofar as it has not already been completed by the proposed sub-contractor, deleting those items which are no longer relevant and

PRE-CONTRACT PRACTICE

        agreeing all the remaining items with the sub-contractor.
- (13) The contractor and the sub-contractor sign Schedule 2 and the contractor signs the tender itself to indicate his acceptance of it, subject only to the architect issuing the nomination instruction.
- (14) The contractor then returns the original, and now completed, Tender NSC/1 and the copies to the architect.
- (15) The architect issues his instruction to the contractor nominating the proposed sub-contractor using Nomination NSC/3, sending with it the original Tender NSC/1.
- (16) At the same time the architect sends to the proposed sub-contractor a copy of Nomination NSC/3, together with a certified copy of the completed Tender NSC/1.
- (17) The contractor and the nominated sub-contractor enter into a Sub-Contract NSC/4, either under hand or under seal, which has been set out previously in Schedule 2 of Tender NSC/1.

Information to be Provided

The information which must be inserted by the architect in Tender NSC/1 before sending it to the proposed sub-contractor with the initial enquiry is set out in detail in Schedule 1 and may be summarised as follows:

- (1) Names and addresses of employer, architect, quantity surveyor and main contractor (if known at the time of the enquiry).
- (2) Particulars of the sub-contract conditions to be used.
- (3) Particulars of the sub-contract fluctuations clause.
- (4) Description of the main contract works.
- (5) Particulars of the main contract conditions of contract and where those conditions may be inspected.
- (6) Whether main contract is to be under hand or under seal.
- (7) How the alternatives in the main contract are dealt with, i.e. whether architect or supervising officer in local authorities edition, alternatives in respect of fire insurance, etc., and whether a provisional sum will be included in the main contract to cover insurance of additional risks in connection with damage to property.
- (8) Particulars of any changes from the printed standard form of main contract conditions.
- (9) Details of all items to be inserted in the Appendix to the main contract conditions.
- (10) Any requirements of the employer affecting the order of the main contract works.

SUB-CONTRACTORS AND SUPPLIERS

(11) Location and type of access to the site.
(12) Particulars of any obligations or restrictions imposed by the employer which are not covered by the main contract conditions.
(13) Any other relevant information.
(14) Date of tender for fluctuation purposes – to be inserted in the appropriate Appendix. The further details normally required if fluctuations are to be calculated using the formula.

By following this procedure it will be seen that the proposed sub-contractor is in possession of full information about the contract from the start and the situation no longer arises where subsequently the main contractor and the sub-contractor are brought together without prior consultation between them. Under the basic method there is full consultation between contractor and sub-contractor prior to nomination and the numerous matters which have hitherto been constant sources of trouble, particularly in relation to the programme of works and special attendance provisions, which are set out in Schedule 2 of NSC/1, are discussed and agreed between contractor and sub-contractor before the architect issues his instruction making the nomination.

In addition, by having the Agreement NSC/2 executed by the sub-contractor and the employer when the Tender NSC/1 is first submitted, the sub-contractor's responsibility direct to the employer with regard to the design of the sub-contract works, and the selection by the sub-contractor of goods and materials in connection with them, is established at the earliest possible moment.

The Alternative Method of Nomination

We turn now to the alternative method of nomination and in this case the Tender NSC/1 and the Nomination NSC/3 are dispensed with. The two documents used in the alternative method are:

- Agreement NSC/2 adapted for use where Tender NSC/1 has not been used (Agreement NSC/2a).
- Sub-Contract NSC/4 adapted for use where Tender NSC/1, Agreement NSC/2 and Nomination NSC/3 have not been used (Sub-Contract NSC/4a).

When the alternative method is to be used this must be stated in the bills of quantities or in the architect's instruction under which the nomination is made. The bills, or the instruction, must also state whether or not Agreement NSC/2a will be entered into, this being optional under the alternative method.

In effect the alternative method is analogous to the common practice under earlier forms of the contract whereby the architect obtained a quotation from a sub-contractor and subsequently nominated that sub-contractor with no formal documentation. Under JCT 80, however, even using the alternative method, once the nomination has been made the contractor and the sub-contractor must enter into the Standard Form of Sub-Contract NSC/4a. The main difference between NSC/4 and NSC/4a is that the latter incorporates an Appendix into which must be inserted the various matters which under the basic method are set out in Tender NSC/1. Under the alternative method, however, these matters will not have been agreed between the contractor and the sub-contractor before the nomination is made and the advantages stemming from prior agreement will have been lost.

As far as the use of Agreement NSC/2a, that is the Employer/Sub-Contractor Agreement under the alternative method, is concerned, this will depend on the nature of the sub-contract works. Where the proposed sub-contractor is responsible for design or the selection of materials or goods then clearly the Agreement NSC/2a must be used especially as, under JCT 80 it is expressly stated that the main contractor has no design responsibility in connection with the sub-contract works. In addition Agreement NSC/2a should also be used if the nature of the sub-contract works is such that delay or default by the sub-contractor might disrupt the works as a whole, giving the main contractor the right to seek an extension of time.

Choice of Method

Clearly the choice of the method to be adopted in the making of any sub-contract is a matter which must receive careful consideration during the pre-contract period. Except where proposed sub-contract works are covered by a provisional sum the decision on the method of making each nomination must be taken before the bills of quantities are completed as the method must be stated in the bills.

Broadly speaking it is recommended that the basic method should always be used in respect of the following:

- Any sub-contract in which design information must be obtained from the sub-contractor during the pre-contract period.
- Any sub-contract, the performance of which is likely to affect the progress of the contract as a whole.
- Any sub-contract which requires prior agreement between the main contractor and the sub-contractor on matters of programme, performance and attendance.

Finally it should perhaps be mentioned that both the basic method and the alternative method can be used on the same main contract. Furthermore the architect may substitute the basic method for the alternative method and vice versa during the course of the contract,

SUB-CONTRACTORS AND SUPPLIERS

provided he has not already issued a notice of nomination under the basic method or a nomination instruction under the alternative method. Such a substitution, however, is treated as a variation and the main contractor would be entitled to reimbursement for any loss or expense he might incur as a result of the substitution. It is clearly advisable, therefore, that positive decisions on nominations are made during the pre-contract period.

Nominated Suppliers

A supplier is nominated or deemed to be nominated if:

- A prime cost sum is included in the bills of quantities or in an instruction regarding the expenditure of a provisional sum and if the supplier is named in the bills or the instruction, or is subsequently named by the architect,

or

- In an instruction regarding the expenditure of a provisional sum or in a variation order the architect specifies materials or goods which can only be purchased from one supplier.

If the nomination of a supplier arises from the expenditure of a provisional sum or under a variation order the materials or goods concerned must be made the subject of a prime cost sum.

If a supplier is named in the bills of quantities, but no prime cost sum is included, that supplier would not be a nominated supplier, it being left to the contractor to inform the architect prior to the contract being placed if he considers that the named supplier should be treated as a nominated supplier and a prime cost sum inserted.

It is advisable that decisions on nominated suppliers should be taken if possible during the pre-contract period so that they can be dealt with in the proper manner in the bills of quantities.

The Joint Contracts Tribunal now publish a form of tender for nominated suppliers and, although the use of this is not mandatory under the terms of the contract, it is strongly recommended that it be used as it includes an Appendix which sets out in detail the terms which must appear in the nominated suppliers conditions of sale, these terms now being incorporated in JCT 80.

Obtaining Tenders

The provisions of JCT 80 both as regards nominated sub-contractors and nominated suppliers, though rather complicated on the face of them, should go a long way to eliminating many of the problems which have arisen in the past in this difficult contractual area. The requirements which are mandatory must, of course, be followed by the employer's professional advisers. Those which are optional should be carefully considered and on the whole it would appear to be advisable that the formal procedures which are available should be followed and the permissible short cuts taken only on very minor matters.

If it is desired to select the proposed nominated sub-contractor or supplier by competitive tender this presents no problems. The use of the Standard Form of Nominated Sub-Contract Tender and Agreement (Tender NSC/1) and the use of the Form of Tender by Nominated Supplier will simplify the obtaining of competitive tenders and will ensure that the firms tendering do so on the precise terms required by the contract.

Generally speaking our recommended procedures for obtaining tenders for the main contract apply equally to sub-contracts.

Standard Forms of Tender are, however, not in themselves sufficient to ensure good tenders and it must be borne in mind that accompanying documentation must adequately cover the work involved. This will normally take the form of drawings together with a specification, or a performance specification and where appropriate, a sub-contract bill of quantities. In all these documents the high standards recommended for the main contract documents should be maintained.

One matter which arises from JCT 80 should be noted insofar as it affects bills of quantities and specifications for sub-contract works. These documents must not now include any prime cost sums. The architect has no power to issue instructions regarding the expenditure of prime cost sums in sub-contract documents and if it is desired to nominate within a nominated sub-contract the works, materials or goods concerned must be covered by a provisional sum.

SUB-CONTRACTORS AND SUPPLIERS

Sub-Contractor Design

JCT 80 does not contemplate design on the part of the contractor. Where such design is involved the JCT has special forms of contract which apply, namely, the Standard Form With Contractor's Design and the Standard Form With Quantities modified by the Contractor's Designed Portion Supplement. These contracts, however, do not fall within the scope of this book since we have concentrated on the traditional system of consultancy design and contractors building to that design.

When the main contract excludes contractor design, it necessarily follows that any sub-contract must do likewise and therein lies a common problem, for inevitably some nominated sub-contractors under JCT 80 will be involved in elements of design.

Where such sub-contract design is properly planned and co-ordinated with the other production information available to the contractor, as it should be in terms of 'good practice', there is usually little practical problem. However, when failures occur, or when delay is experienced, the employer is liable to suffer considerable loss in terms of time and/or money and he will have no contractual right to recover his losses.

In recognition of this situation a system of separate agreements has been incorporated in the sub-contract, whereby a direct contractual link is formed between the employer and the sub-contractor.

Agreements NSC/2 and NSC/2a

Under JCT 80 the Employer/Nominated Sub-Contractor Agreement, NSC/2, is an integral part of the basic method of nomination. Under the alternative method the agreement is optional and when required, the modified form NSC/2a is used (NSC/1 having been dispensed with).

Under NSC/2, in consideration of nomination, the sub-contractor warrants to exercise all reasonable skill and care in:

- his design of the sub-contract works.
- his selection of materials and goods for the sub-contract works.
- satisfying any performance specification requirements.

and will:

- supply all information and drawings in time (so as not to give cause for a claim for delay).
- not default on his obligations under the sub-contract.
- not cause delay (so as to give the main contractor cause to claim an extension of time).
- indemnify the employer against loss arising from re-nomination if he defaults.

PRE-CONTRACT PRACTICE

On the other side, the employer warrants that:

- the architect will direct the contractor and inform the sub-contractor as to the amounts certified in interim and final certificates.
- he will operate the system of final payment of sub-contractors laid down in the main contract.
- he will operate the provisions for direct payment laid down in the main contract.

The agreement also contains arbitration provisions. It is signed or sealed by the parties.

NSC/2a is essentially the same in respect of these various rights and duties. It is merely redrafted, and renumbered to take account of the absence of NSC/1.

Chapter 8

OBTAINING TENDERS

Selective Tendering

We referred in the Introduction to this book to the changes which are taking place in procedures in the building industry; changes necessitated by the increasing use of specialised methods of construction where the contractor's participation is required in the design stage, by the use of larger factory-made components and building systems, or by heavy pressure from clients to reduce the pre-contract period so as to let the design and construction programmes overlap.

All these circumstances involve the early appointment of the contractor and this calls for modification of traditional tendering procedures. We have dealt with these extensively in our book *Tenders and Contracts for Building*. However, traditional methods are still the most widely used and we set out in this chapter what we consider to be good procedure for these cases. To some extent they can be modified in special cases. The principles to be followed are clearly and logically defined in the Code of Procedure for Single Stage Selective Tendering 1977, published by the National Joint Consultative Committee for Building in collaboration with the Department of the Environment and the Joint Committees for Scotland and Northern Ireland.

We strongly recommend that open tendering should be avoided and that tenders should normally be obtained from a selected list of contractors. The list should be discussed with the client about three months before tenders are required. The selected list may be drawn up from firms known to the architect, quantity surveyor or client, or by selection from firms who apply in response to advertisements in the technical and local press. These advertisements should indicate the size, nature and location of the job and the date the tender documents will be ready.

It should perhaps be mentioned here that under the rules of the European Economic Community contracts for public works exceeding a certain threshold value (excluding VAT and the value of nominated sub-contracts), must be advertised in the *Official Journal of the European Communities* before tenders are invited. It is still permissible to compile a selected list of contractors from those who respond to the advertisement and to that extent selective tendering in these circumstances is not affected. The rules are set out in EEC Directives 71/305, 72/277 and 78/669, and DOE Circular 67/78.

The threshold value (which is laid down as 1 million European units of Account) is determined every two years by reference to average national currency conversion factors. From 1st January 1986 the sterling value was set at approximately £570,000 and a further determination will operate from January 1988, 1990 and so on.

The number of firms to be invited to tender should, of course, depend on the size and type of contract. The Code of Procedure for Single Stage Selective Tendering 1977 recommended the following as a guide in relation to building cost at that time:

Size of Contract	Maximum number of tenderers
Up to £50,000	5
Between £50,000 and £250,000	6
Between £250,000 and £1 million	8
£1 million plus	6

We would suggest that today, for contracts up to £100,000, four or five tenderers are quite sufficient and for contracts of greater value, six should be the maximum.

The cost of tendering is high and the larger the tender lists become the greater will be the cost of abortive tendering and this must be reflected in building prices. When the list has been settled one or two names should be appended in order that they may replace any firms on the list that do not accept the invitation.

The firms selected should be of similar size and standard and should be known to be suitable for the particular job.

Preliminary Enquiry

About a month before the tender documents are sent out, a letter should be sent to the selected tenderers asking them whether they wish to tender and giving the following information:

  (1) Job . . . . .
  (2) Building owner . . . . .
  (3) Architect . . . . .
  (4) Quantity surveyor . . . . .
  (5) Engineering consultants . . . . .
    . . . . .
  (6) Location of site . . . . .

OBTAINING TENDERS

(7) General description of works . . . . .
(8) Approximate cost range £ – . . to £ – . .
(9) (a) Form of Contract to be used . . . . .
 (b) Clause 38, 39 or 40 to apply . . . . .
(10) Anticipated date of possession . . . . .
(11) Indication of clients' requirements for programme for works and date of completion . . . . .
(12) Value of Bond . . . . . % of contract sum (if required)
(13) Approximate date for dispatch of all tender documents . . . . .
(14) Tender to be returned by . . . . .

At the same time, any contractors who asked to tender in reply to an advertisement and who were not included in the selected list should be so informed.

Tender Period

Contractors must be allowed ample time for tendering if realistic tenders are to be obtained. The time required by a contractor to prepare a tender depends not only on the size of the job, but also on its complexity, and varies considerably.

We consider that as a general rule at least four weeks should be allowed for tendering, though in some cases for small simple work this might be reduced to three weeks, and in other cases more time will be needed. It is imperative that tenderers shall have sufficient time to prepare sound and proper tenders, including, where necessary, prices from specialist sub-contractors.

PRE-CONTRACT PRACTICE

Tender Documents

The following documents should be sent to the selected tenderers:

(1) Two unbound copies of the bills of quantities.
(2) Drawings in accordance with the requirements of SMM6 (see Example O at the end of this chapter).
(3) Two copies of the Form of Tender.
(4) An addressed envelope for the return of the tender suitably marked on both sides with the word 'Tender' and marked on the front with the name of the job and the time and date to be delivered.
(5) If the priced bills of quantities are to be returned at the time of the tender, an addressed envelope to contain the bills, marked with the name of the job and with a space for the contractor's name on the outside.

We strongly recommend that, if it is proposed to use the method of dealing with errors in tenders as outlined in Alternative 2 of Section 6 of the Code of Procedure for Single Stage Selective Tendering, priced bills of quantities must be returned with the tender.

In the letter accompanying these documents the following information and instructions should be given:

(1) Date and time tenders are to be returned.
(2) Place where all other drawings may be inspected and with whom arrangements have to be made for the purchase of additional copies of special drawings if required by specialists.
(3) Arrangements for inspection of the site.
(4) Time and place of opening of tenders and whether contractors may be present.
(5) When priced bills of quantities are to be submitted at the time of tendering, assurances that these will not be opened unless the contractor's tender is under consideration and that priced bills will be returned to unsuccessful contractors.
(6) Method of correction of priced bills (Section 6 of the NJCC Code of Procedure for Single Stage Selective Tendering 1977) Alternative 1/Alternative 2 to apply.
(7) Instruction that contractors are to acknowledge receipt of all documents.

Receipt of Tenders

When tenders are received they should in no circumstances be opened before the proper time.

Although it is customary to state that no pledge is given to accept the

OBTAINING TENDERS

lowest or any tender, we consider that, unless special circumstances arise, the most advantageous tender should be accepted.

The selection having been made, the quantity surveyors should, as quickly as possible, report on the following:

(1) The checking of the arithmetic in the priced bills of the prospective contractor and ensuring that any amendments notified during the tendering period have been made.
(2) Inspection of prices in the bills and advising whether they are fair and reasonable noting any abnormal prices.
(3) Checking that the tenderers' basic rates, if any, are reasonable.
(4) Making any analysis of the tender considered necessary.
(5) Comparison with cost target and opinion on tender.
(6) Recommendation for acceptance or otherwise.

Correction of Errors in Tenders

In the event of a serious error of pricing or arithmetic in the tenderer's bills of quantities, this should be dealt with in accordance with the alternative stated in the formal invitation to tender.

Under alternative 1 the tenderer should be given details of errors in the bills and afforded an opportunity of confirming or withdrawing his offer. If the tenderer withdraws, the priced bills of the second lowest should be examined.

In alternative 2 the tenderer should be given an opportunity of confirming his offer or of amending it to correct genuine errors. Should he elect to amend his offer and the revised tender is no longer the lowest, then professional judgement must be exercised in determining which other tenders are to be examined in detail.

Under either alternative, where the tenderer confirms his offer, an endorsement should be added to the priced bills indicating that all rates or prices (excluding preliminary items, contingencies, prime cost and provisional sums) inserted therein by the tenderer are to be considered as reduced or increased in the same proportion as the corrected total of priced items exceeds or falls short of the original total of such items.

Alternatively, there may be cases where it is appropriate to make the adjustment in the preliminaries, or profit if separately stated, and this is certainly a simpler method saving much calculation later.

In either case the contractor's agreement must be obtained and the endorsement signed by both parties.

If under the second alternative the tenderer elects to amend his tender figure, and possibly certain of the rates in his bills, he should either be allowed access to his original tender to insert the correct details and to initial them, or be required to confirm all the alterations in a letter. If in the latter case his revised tender is eventually accepted, the letter should be conjoined with the acceptance and the amended tender figure and rates in it substituted for those in the original tender.

Report on Tenders and Acceptance

The architect as well as the quantity surveyor may wish to comment and report on the tenders received and both reports should go to the employer with a recommendation to assist him in making his final decision on the tender to be accepted.

Not more than four weeks after tenders have been received, the employer should have either accepted a tender or rejected all tenders, or accepted a tender subject to modifications to be agreed.

A contract should be drawn up and signed as soon as possible and in any case before possession of the site is given. The date for possession should either be established in the tender documents, or at latest when the tender is accepted.

Notification of Tender Results

Once the contract has been let, each tenderer, including the successful one, should be supplied with a list of the tender prices. In this way the tenderer's actual figure is not disclosed whilst each tenderer knowing his own figure will see how he stood in relation to the rest.

The principal advantages of this established tendering procedure are that it is the most effective way of letting a contract on a competitive financial basis, and at the same time it establishes for the employer an accurate indication of his financial commitment. However, both these advantages can be largely maintained when other methods of obtaining tenders are adopted.

Competition of varying degrees can be introduced in the early appointment of the contractor and financial control can be maintained by making full use of up-to-date cost control techniques.

To deal in detail with the many variations in tendering methods is not practical in a book of this size and scope and we have therefore published another work, *Tenders and Contracts for Building*, to cover the subject of the choice of appropriate tendering and contractual arrangements in various circumstances.

Bonds (Guarantee of Performance)

The employer may require the contractor to provide a bond for the due performance of the contract. Such a bond will normally be obtained by the contractor from an insurance company or a bank, or alternatively, where the contractor is a subsidiary company of a larger organisation, it may take the form of a guarantee from the parent company. The bond holder or guarantor and the terms of the bond must, of course, be approved by the employer and the amount or surety provided will normally be 10% of the contract sum, this amount becoming available to the employer to meet any additional expense he incurs as a result of the contractor failing to execute the contract or otherwise being in breach of his obligations under it.

As the raising of a bond can be a financial burden on the contractor in terms of his borrowing etc., it is considered good practice that the bond should be released on reaching practical completion.

Whether or not a bond is required is one of those matters which must be settled before the documents are sent out to tender so that the contractor, who is responsible for all costs in connection with the bond, can include those costs in his price.

There is no standard form of bond published specifically for building works, but most bonds are similar in form and we reproduce in *Contract Administration* the standard form incorporated in the Institution of Civil Engineers (ICE) Conditions of Contract and an example of a bond in the form of a parent company guarantee.

EXAMPLE O: LIST OF DRAWN AND OTHER INFORMATION TO BE PROVIDED IN SUPPORT OF THE BILLS OF QUANTITIES*

(1) Location and Supplementary Drawings and Specifications

    (A) Block Plan – to identify site and outline of building in relation to Town Plan or other wider context.
    (B) Site Plan – to locate position of the buildings in relation to setting out points, means of access and general layout of site.
    (C) General Location Drawings – to show the positions occupied by the various spaces in a building and general construction and location of principal elements.
    (D) Specifications – for plumbing and mechanical engineering and electrical installations.

Details of general and supplementary location drawings and specifications required in connection with the various SMM work sections

Note: Unless otherwise indicated, only General Location Drawings (as defined in (C) above) are required for the stated Work Section.
    (i)   Demolition.
    (ii)   Excavation and Earthwork.

*Reproduced from Dearle and Henderson (1979) *SMM6 Handbook*. London: Granada Publishing.

- (iii) Piling and Diaphragm Walling. Location drawings to show:
  - (a) The general piling layout.
  - (b) The positions of any different types of piles.
  - (c) The positions of the piling work within the site and of existing services.
  - (d) The arrangement of diaphragm walls and their relationship to surrounding buildings.
  - (e) The depths, lengths and thicknesses of diaphragm walls.
- (iv) Concrete Work. The Location drawings and, for concrete framed structures and concrete to steel framed structures, further drawings as may be required to show:
  - (a) The relative positions of differing types of construction.
  - (b) The size of principal structural members including thickness of floor slabs.
  - (c) The permissible loads on slabs and beams which may carry temporary supports of framework relative to the time elapsed since the slab was cast. Alternatively may be given on a schedule.
- (v) Brickwork and Blockwork. The Location or further drawings as may be required to show:
  - (a) Plan of each floor level giving the positions of and materials to be used in all walls and partitions.
  - (b) All external elevations showing the formation level, floor levels and materials used.
- (vi) Underpinning. The Location or further drawings to show the location and extent of the work and particulars of the existing structure to be underpinned.
- (vii) Rubble Walling. As (v).
- (viii) Masonry. As (v).
- (ix) Asphalt Work. The Location or further drawings as may be necessary to show:
  - (a) Plan of each level indicating the extent of the work and its height above ground level together with any restrictions on the siting of pots.
  - (b) A section indicating the extent of any tanking.
- (x) Roofing. The Location or further drawings to show the extent of the work and its height above ground level.
- (xi) Structural Steelwork. The Location drawings and, for fabricated steelwork, further drawings to show:
  - (a) The position of the work in relation to other parts of the work and of the proposed buildings.
  - (b) The types and sizes of structural steel members and their positions in relation to each other.
  - (c) Details of connections or particulars of the reactions, moments and axial loads at connection points.

OBTAINING TENDERS

(xii) Plumbing and Mechanical Engineering Installations. The Location drawings together with detailed specifications of mechanical engineering installations and drawings indicating the scope of the work.

Note: The result of the above will mean that for fabricated steelwork full framing plans should be provided for measurement purposes.

(xiii) Electrical Installations. The Location drawings (which are to include final sub-circuits showing layout of points) together with detailed specifications of the electrical installations and drawings indicating the scope of the work.
(xiv) Floor, Wall and Ceiling Finishings. The Location drawings to show:
   (a) Dry linings and partitions and suspended ceilings and support work except in the cases of rectangular linings, partitions and ceilings without integrated services.
   (b) Carpeting where width or direction in which it is to be laid is specified.
(xv) Glazing.
(xvi) Painting and Decorating.
(xvii) Drainage. The Location or further drawings to show layout of the drainage.
(xviii) Fencing.

(2) Component Detail Drawings

Drawings to show all the information necessary for the manufacture and assembly of the component.

Component detail drawings are specifically required for the following

Masonry:
(a) Carving and Sculpture.
Woodwork:
(a) Staircases and short flights of steps.
(b) Balustrades (if not included with staircases).
(c) Fittings.
Metalwork, Composite items e.g.:
(a) Windows and doors complete with frames, mullions, transoms, hinges and fastenings.
(b) Rooflights, laylights and pavement lights.
(c) Balustrades and staircases.
(d) Duct covers and frames.
(e) Gates, shutters and the like complete with frames, guides, tracks, etc. indicating approximate weight.

(f) Cloakroom fittings, cycle racks, storage racks and the like.
(g) Grilles and gratings.
(h) Ladders.

INDEX

architect, 13, 14, 16, 19, 25, 26, 33, 35, 42, 50, 51, 63
Architect's Job Book (RIBA), 6

bills of quantities, 34, 40, 43, 50, 67–77, 84, 85, 86
  annotated, 69, 70
  basic information, 67
  drawings upon which based, 39
  elemental, 69
  examples of, 72–7
  list of drawings and other information to be provided in support of, 95–8
  measured works, 68
  operational, 69, 70
  preambles to, 68
  preliminaries, 68
  role of, 67
bond (guarantee of performance), 94
British Standards, Specifications and Codes of Practice, 63
  BS 1192: 1984, 31, 32, 44
  BS 3429, 31
budget, 16
builder, 33, 42, 50, 51 *see also*, contractor
Building Cost Information Service of the RICS, 19, 26, 70
  standard form of cost analysis, 19
building process, 35
building owner, 13, 14
  see also, employer

certificates, interim, 88
  final, 88
claims for extras, 35
Code of Procedure for Single Stage Selective Tendering, 2, 89, 90, 92
computer aided design (CAD), 35
computers, 27
  costing systems, 27
  hardware, 27
  software, 27

Contract Administration (The Aqua Group), 2, 95
contractor, 31, 32, 34, 35, 51
  see also, builder
Co-ordinating Committee for Project Information (CCPI), 40
cost, 13
  amplified cost plan sheets, 26
  checking, 25
  control, 13, 14, 16
  in-use, 13
  limit, 14, 15, 27
  plan, 13, 14, 18, 19, 25, 26
  planning, 13, 14, 16, 19, 69
  trends, 27
  unit limits, 14

design, brief, 6
  detail stage, 19
  process, 14
  team, 11
designers, 25, 31
DOE Circular, 67/78, 90
domestic sub-contractors, 79
drawings, 'as-built', 40
  contents, 35–9
  for records, 39, 40
  scale of, 32, 33

elemental, breakdown, 26
  cost summary, 26
elements, list of, 19–25
  of the building, 13, 19
employers, 14, 60
  brief, 15
  financial commitment, 34
  intention, 68
  money, 19
estimates, approximate, 13, 14, 16, 19
  based on approximate quantities, 17, 18
  based on floor area, 16
  information required for, 17, 18

INDEX

preliminary, 16
European Economic Community, 89
  contracts for public works, 89
  EEC Directives, 90
  Official Journal of, 89

final account, 25
funding arrangements, 7

Handbook of Architectural Practice
  (RIBA), 5

ICE conditions of contract, 95
interim certificates, valuations for, 67

Joint Contracts Tribunal (JCT), 2, 11, 85
  Agreement for Minor Building Works, 11
  Intermediate Form of Building Contract 1984 (IFC 84), 11, 79
  Standard Form of Building Contract 1980 (JCT 80), 11, 51, 85, 86
  Standard Form of Building Contract with Contractor's Design 1981, (CD/81), 11
  Standard Form of Nominated Sub-contract Tender and Agreement, 69, 80, 81, 82, 83, 84, 86, 87, 88

National Building Specifications (NBS), 63
National Joint Consultative Committee for Building, 2
nominated sub-contractors, 68, 78–88
  Employer/Sub-contractor Agreement, 79
nominated suppliers, 68, 85, 86
  form of tender, 86

prime cost sums, 68, 80, 85, 86
private developer, 16
provisional sums, 69, 80, 84, 85

quantity surveyor *see*, surveyor, quantity

schedules, 31, 34, 40, 42
  characteristics, 42, 43, 44
  examples of, 45–9
  item, 42
  size, 42, 44
Simon Report, 3
sketch, scheme, 12
  plans, 18
specification, 50
  examples of 64–6
  for sub-contract works, 86
  headings for, 52–63
Standard Form of Building Contract *see under*, Joint Contracts Tribunal
Standard Form of Nominated Sub-contract Tender and Agreement *see under*, Joint Contracts Tribunal
Standard Method of Measurement, 34, 40, 68, 92
statutory authorities, 68
sub-contractor design, 87
surveyor, quantity, 13, 16, 17, 25, 26, 31, 32, 33, 34, 42, 50, 51, 63

tenders, competitive, 86
  correction of errors in, 93
  documents, 26, 92
  form of, 92
  notification of results of, 94
  open, 89
  period, 90
  receipt of, 92
  report on, 94
  selective, 89
  standard forms of, 86
Tenders and Contracts for Building (The Aqua Group), 2, 89, 94

variation order, 80
VAT, 12, 18

word processors, 63
working drawings, 18